6683d

Evolutionary Economics

Evolutionary Economics

Post-Schumpeterian Contributions

Esben Sloth Andersen

Pinter Publishers
London and New York

Distributed in the United States and Canada by
St. Martin's Press

Pinter Publishers Ltd.
25 Floral Street, London WC2E 9DS, United Kingdom

First published in 1994

© Esben Sloth Andersen, 1994

Apart from any fair dealing for the purpose of research or private study, or criticism or review, as permitted under the Copyright, Designs and Patents Act, 1988, this publication may not be reproduced, stored or transmitted, in any form or by any means or process without the prior permission in writing of the copyright holders or their agents. Except for reproduction in accordance with the terms of licences issued by the Copyright Licensing Agency, photocopying of whole or part of this publication without the prior permission of the copyright holders or their agents in single or multiple copies whether for gain or not is illegal and expressly forbidden. Please direct all enquiries concerning copyright to the publishers at the address above.

Esben Sloth Andersen is hereby identified as the author of this work as provided under Section 77 of the Copyright, Designs and Patents Act, 1988.

Distributed exclusively in the USA and Canada by St. Martin's Press, Inc., Room 400, 175 Fifth Avenue, New York, NY10010, USA

British Library Cataloguing in Publication Data

A CIP catalogue record for this book is available from the British Library

ISBN 1 85567 042 9

Library of Congress Cataloguing-in-Publication Data

Andersen, Esben Sloth.
    Evolutionary economics: post-Schumpeterian contributions / Esben Sloth Andersen.
        p.   cm.
    Includes bibliographical references and index.
    ISBN 1-85567-042-9
    1. Evolutionary economics. 2. Technological innovations. 3. Diffusion of innovations. 4. Competition. I. Title
HB97.3.A53 1994
330.1—dc20   93-39415
                                                                                                                                CIP

Printed and bound in Great Britain by
Biddles Ltd., Guildford and King's Lynn

Contents

Preface ix

1. Approaching economic evolution 1

1.1. *The contours of an evolutionary research programme around 1950* 1
    1.1.1. The empirical orientation as Schumpeter's final thesis 2
    1.1.2. The algorithmic approach and Goodwin's prologue 5
    1.1.3. The population perspective as introduced by Alchian 9

1.2. *Surveying evolutionary economics* 13
    1.2.1. Crude definitions 13
    1.2.2. On the modern history of evolutionary economics 16

Notes 21

2. Schemes of punctuated evolution and jerky innovation 26

2.1. *From railroadization to analytical schemes* 26

2.2. *Measuring the application of routines* 31

2.3. *The scheme of punctuated evolution* 36

2.4. *Punctuated evolution and Schumpeter's 1910-theses* 40

2.5. *Economic paradigms and evolution* 44

2.6. *Two principles of interface design* 49
    2.6.1. The principle of commodity abstraction 49
    2.6.2. The principle of interactive learning in product innovation 52

2.7. *Conclusions on paradigms* 55

Notes 59

3. Density-dependent diffusion and innovation 63

3.1. *The network of interrelations* 63

3.1.1. Integrating 'ecological' and evolutionary analysis — 63
3.1.2. The unfinished growth pole analysis as an example — 65

3.2. *Logistic diffusion in post-Schumpeterian terms* — 67

3.2.1. Introduction — 67
3.2.2. The logistic difference equation and diffusion of innovations — 69
3.2.3. Short and long innovation-based business cycles — 72
3.2.4. Some limitations of the simple logistic-curve approach — 76

3.3. *Lotka-Volterra inspired models of interdependent diffusion?* — 78

3.3.1. Fables of density-dependent evolution — 78
3.3.2. The interdependence of applications of different innovations — 81
3.3.3. Evolutionary interaction and niches — 84

3.4. *The logistic curve and economic decision-making* — 87

3.4.1. Firm-level analysis — 87
3.4.2. Pioneering and crowding modes of behaviour — 88

Notes — 91

4. Exploring the process of 'Schumpeterian competition' — 95

4.1. *The complex world* — 95

4.1.1. Introduction — 95
4.1.2. Problem complexity and Artificial Intelligence — 96

4.2. *The pioneering work of Nelson and Winter* — 101

4.2.1. An evolutionary synthesis — 101
4.2.2. Typical Nelson-and-Winter simulation models — 104

4.3. *A formal version of the computational structure of typical Nelson-and-Winter models* — 109

4.3.1. Inheritance from period $t-1$ — 110
4.3.2. Short-run behaviour — 110
4.3.3. Productivity change through innovation and imitation — 111
4.3.4. Change in physical capital — 112
4.3.5. State of the system at the beginning of period $t+1$ — 113
4.3.6. Implementing the formal specifications — 114

4.4. *The extendibility of Nelson-and-Winter models* — 118

4.4.1. Satisficing behaviour — 118
4.4.2. Exit and entry — 119
4.4.3. Industry creation — 123
4.4.4. Product quality, product innovation and imitation — 124

4.5. *Appreciating Nelson and Winter's modelling scheme* — 131

Notes — 138

5. The evolution of strategies of transaction: an algorithmic story 143

5.1. *Rule change in iterated computations* 143

5.1.1. Rule-based systems and genetic algorithms 143
5.1.2. Arthur's application of classifier systems 147
5.1.2. From production routines to rules of interaction 150

5.2. *From Axelrod to the Trader's Dilemma* 152

5.2.1. Axelrod's evolution of cooperation 152
5.2.2. A fable of the difficult evolution of cooperation 155

5.3. *Iterated games during a trading period with fixed rules* 159

5.3.1. The possibility of retrospection 159
5.3.2. Strategies for the iterated Trader's Dilemma 161
5.3.3. Complexity and bounded rationality 164
5.3.4. The conditional values of strategies 166
5.3.5. The overall value of a strategy for a month's trading 167

5.4. *The evolutionary change of strategies* 169

5.4.1. The changing frequency of the different strategies 170
5.4.2. The theoretical fitness of strategies 172
5.4.3. The modified genetic algorithm 174
5.4.4. The multi-patterned evolutionary dynamics 176

Notes 180

6. Research horizons 185

6.1. *Towards an evolutionary-economic synthesis* 185

6.2. *The diversity of evolutionary explanations* 190

6.2.1. The past multitude of biological explanations 190
6.2.2. The diversity of evolutionary-economic explanations 193

Notes 196

Appendix: Algorithmic notation and the programming of Nelson-and-Winter models 198

A.1. *Algorithmic modes of describing evolutionary models* 198

A.1.1. The notation of this book 198
A.1.2. The analytical cycle 199
A.1.3. The MAPLE notation 201

A.2. *Nelson-and-Winter models formulated as MAPLE programs* 205

A.2.1. Data structures 205
A.2.2. The recursive master procedure 209

A.2.3. State initialisation	209
A.2.4. The general computation of the state variables	210
A.2.5. The short-term market process	212
A.2.6. Finding the new technology of a firm	212
A.2.7. The firm's investment and depreciation process	214
A.2.8. Collection of statistical data	216

A.3. *Starting an object-oriented version of Nelson-and-Winter models*   216

Notes   219

References   220

Index   232

Preface

This book concentrates on core elements of the new evolutionary economics, especially the elements covered by what may be called Artificial Economic Evolution. But it also relates to the more general revival of studies on economic and technological evolution, and thus to the fact that many researchers today have accepted Schumpeter's (1942/87, 82) proposition that the 'essential point to grasp is that in dealing with capitalism we are dealing with an evolutionary process.' The difficult tasks have been to visualise and analyse such a process, to move from the mainstream studies of 'how capitalism administers existing structures' to the more difficult studies about 'how it creates and destroys them.' (p. 84) In performing these tasks modern economists have access to analytical tools that were non-existent when Schumpeter developed his 'magnificent dynamics' (Baumol). We now have the possibility of treating complex evolutionary processes with an increasing degree of clarity and rigour. At the same time we have to recognise a danger that Schumpeter's insights are ignored just as are the works of other pre-war contributors to the study of economic evolution. There is also a danger that the evolutionary-oriented studies of some economic historians like Chandler disappear from the analytic horizon of evolutionary-economic researchers.

To avoid the crowding out of evolutionary economics, this book presents major tools and results of new evolutionary economics in the context of a larger evolutionary-economic research programme which can more or less clearly be found in Schumpeter's work. In such a context, the present ability to cope with evolutionary processes is fully recognised. However, the ability to synthesise mechanisms concerning the creation, transmission, and selection of behavioural rules into computer-based studies is only a very first step on the way towards a real understanding of the processes of capitalist and post-capitalist evolution. There is also a need for a second synthesis, a synthesis between broad and descriptive accounts of economic transformation and the clear-cut analysis of artificially limited evolutionary processes. The book tries to demonstrate that a viable new evolutionary economics may be defined in terms of these two types of syntheses.

The scope of this book is quite broad and it exploits different sides of my previous work in such diverse fields as economics, biology, and computer science, which all have influenced the way in which I see evolutionary economics. More or less by accident, my diverse interests appear to be rather important to research work in evolutionary economics but the reader should, of course, be aware of the possible biases in my approach to the area. Even within the economic literature there are clear biases which mainly spring from the attempt to create a kind of 'dialogue' between Schumpeter and new evolutionary economics. However, even within my chosen area, I have had to make severe delimitations. One of the remaining tasks is to rethink important parts of industrial economics or industrial organisation from Bain (1956) to Tirole (1988), a body of literature which is beyond the limits of the present book but within the realm of important inspirations to evolutionary economics.

The introductory chapter 1 concerns to a large extent the broader issues which have been suggested above; at the same time it gives a condensed progress report on evolutionary-economic analysis. Much of the exposition deals with what I consider to be four major characteristics of a viable new evolutionary economics: population thinking, empirical orientation, an algorithmic approach, and a relationship to old evolutionary economists (like Schumpeter). In chapter 2 I turn to a reconstruction of Schumpeter's schemes of routine and innovation. Here I deal with the idealised measurement of the application of routines which leads to precise definitions of evolution and non-evolution. These definitions are used to describe Schumpeter's overall scheme of what may be called 'punctuated evolution', i.e. an evolutionary process which once in a while is punctuated by the emergence of relative stasis in the economic routine system. The individual routines and the jump-wise renewal of them which is presupposed by Schumpeter is then dealt with in terms of 'economic or techno-economic paradigms'. I suggest that the major source of stability in the routine system should not be sought within the individual economic units but in the interfaces between them, defined by their repeated transactions. All in all, the chapter emphasises a number of areas of central relevance to Schumpeterian thought and analysis which have not yet been explored by modern evolutionary economics.

The three following chapters deal with modelling work performed around and within new evolutionary economics; especially they deal with the development of the algorithmic approach of Artificial Economic Evolution (an expression which suggests some positive and negative analogies and relationships *vis-à-vis* Artificial Intelligence). In chapter 3 it is demonstrated how the modern studies can help to explore many of the patterns revealed by Schumpeter's analytical scheme as well as his case of 'railroadization'. Especially, it is shown how the S-curve approach to the study of the diffusion of innovations can be extended to deal with the 'ecological' interaction between the application of different routines. Furthermore, some ways of using the S-curve to explore pioneering and crowding strategies are suggested. In chapter 4 Nelson and Winter's early and important studies of evolutionary-economic processes by means of computer models are dealt with. These models helped to make a synthesis between the theories of rule-based behaviour, the old Schumpeter's ideas of the innovative activities of firms, and the ordinary economic account of the market as a selection mechanism. In this way they were able to simulate processes in which innovators and imitators interacted in a process of 'Schumpeterian competition'. In chapter 5 more abstract evolutionary processes emerging from an iterated Prisoner's Dilemma game with limited information are dealt with. The experimentally oriented approach developed by Axelrod in his analysis of the evolution of cooperation is emphasised and the introduction of new variety into the game is explored in terms of the so-called genetic algorithms. Some new aspects of this area of studies emerge from a reinterpretation of the game as a Trader's Dilemma. The general applicability of such a kind of Artificial Economic Evolution is emphasised in a final section on rule-based systems and genetic algorithms. Chapter 6 contains a short statement of the proposed research programme which is embodied in the structure of the book, as well as in its ambition to promote a modern evolutionary-economic synthesis.

The different chapters of the book do not cover the full scale and scope of recent contributions to the rather confusing area of evolutionary economics (see, e.g., Saviotti and Metcalfe, 1991; Witt, 1991b; 1992a, 1993; Foray and Freeman, 1993; Day and Chen, 1993; contributions to the *Journal of Evolutionary Economics*, 1991 ff.). To me it has been more important by means of the algorithmic approach to present and rethink a set of near-classic themes and models. Through an exposition, which relates to, e.g., Schumpeter, Goodwin, Nelson & Winter, and Axelrod, I hope to help new researchers to enter the field of evolutionary economics without an immediate lock-in to a specific modelling trajectory. I believe that such an exposition of major types of models will not only encourage the reader to develop these classic contributions but also to confront alternative proposals. The bias of the book is rather related to its algorithmic approach which may appear to encourage researchers to neglect a formalised, deductive approach to evolutionary processes. But the bias is the result of a necessary specialisation of the present book, and does not reflect a general advise to jump from standard-mathematical to computer-science analysis. I believe that evolutionary economics cannot be developed broadly without the interaction of formal and algorithmic approaches. In this respect, the book cannot stand alone but should be supplemented with works which emphasise a standard mathematical approach to evolutionary economics (see the literature listed above).

There is no chapter with 'policy conclusions'. Policy conclusions should not be made at the beginning of a research programme but as an outcome of its results. However, there is so much confusion about the policy relevance of evolutionary economics that a few remarks are relevant: Often it is said that evolutionary analysis gives no policy conclusions other than *laisser faire*. At other times it is suggested that this conclusion should be supplemented with harsh conservatism. This is not necessarily so! Evolutionary analysis may help the most powerful firms as well as the 'countervailing forces'; it may enlighten supply-side economics as well as post-Keynesian employment policies; it may be used in the competition between national systems of innovation but also in a new type of development policy which includes environmental concerns. These statements have necessarily a postulating character. Let me, therefore, quote the opinion of the central authority in the field. In his preface to *Business Cycles* Schumpeter emphasised:

> What our time needs most and lacks most is the understanding of the process which people are passionately resolved to control. To supply this understanding is to implement that resolve and to rationalize it. This is the only service the scientific worker is, as such, qualified to render. As soon as it is rendered everyone can draw for himself the practical conclusions appropriate to his individual interests or ideals. And it will be seen ... that my analysis can in fact be used to derive practical conclusions of the most conservative as well as the most radical complexion, exactly as one and the same body of engineering or medical knowledge can be used for the most varied purposes. [/] But scientific analysis of an organic process easily creates the impression that the analyst 'advocates' letting the process alone. ... [However,] my analysis lends no support to any general principle of *laisser faire* ... (Schumpeter, 1939, vi)

These words are well-taken. Even if an analysis often creates the impression that the author 'advocates' the evolutionary tendencies which are treated. However, no such connection exists. Actually, there is a huge gap between the Scandinavian-type context of the present author and the Hapsburg-empire setting in which Schumpeter developed many of his views.

Now to a few practical issues: The reference system of the book emphasises collections of papers in order to ease the reader's way through the very heterogeneous literature. However, the dating of the articles is important for several purposes. Therefore, Alchian (1950/93) refers to a paper from 1950 reprinted in Witt (ed.), *Evolutionary Economics*, 1993; similarly, Winter (1984/91) is found in Wood (ed.), *J.A. Schumpeter: Critical Assessments*, 4 vols., 1991. I also use this method of citation for indicating the date of both the original edition and my edition of a monograph.

The present book does not include all the material produced in my related research projects on Schumpeter and evolutionary economics. In respect to this research programme the present work on 'Evolutionary Economics: Post-Schumpeterian Contributions' represents a second part. It can be thought of as preceeded by a 'part I' which includes studies which interpret Schumpeter as a special-type evolutionary economist (including Andersen, 1991b, 1991c, 1991d, 1991e, 1992b, 1993a, 1993b, 1993d). It is followed by a metaphorical 'part III' which includes studies which apply evolutionary-economic concepts and schemes of analysis to such areas as product innovation and quality control as well as national systems of innovation (Andersen, 1991a, 1992a, 1993c). To the extent that these materials have not yet been transformed into publications, they are available for interested researchers.

Some of the work presented in the present book was made possible by a one-year senior scholarship financed by the University of Aalborg. For the rest of the period I have used parts of my research time as an associate professor (and parts of my evenings and nights). The work has been associated with the research work of the IKE Group on the economics of industrial, technological and institutional change, Department of Business Studies (earlier: Institute of Production), University of Aalborg as well as to the PhD programme in Technology Policy, Innovation and Socio-Economic Development, Department of Economics and Planning, University of Roskilde.

In the development of different parts of the book I have had helpful comments from or discussions with many researchers, although I have the sole responsibility for the final result. I especially thank Jan Annerstedt, Jerome Davis, Claus Emmeche, Jan Fagerberg, Chris Freeman, Ole Hyldtoft, Björn Johnson, Christian Knudsen, Kristian Lindgren (permission to use figures 5.2 and 5.3), Bengt-Åke Lundvall, Stan Metcalfe, Erik Mosekilde, Jørgen Østergaard, Keith Pavitt, Jørgen Lindgaard Pedersen, Georg von Wangenheim, Ulrich Witt, and the PhD students in economics and technical change at the Universities of Roskilde and Aalborg. Dorte Køster has helped me to organise the project. Maureen McKelvey has corrected my English in most of the book, and she has also helped to sharpen up several arguments. Similar help with language and arguments was given by my father, Bent Andersen, who died before the book was completed. The book is dedicated to his memory.

1. Approaching economic evolution

1.1. The contours of an evolutionary research programme around 1950

The study of economic evolution presupposes a Copernican turn of perspective *vis-à-vis* standard forms of economic analysis. Instead of dealing with an economic system which is only adapting to exogenous change, we study an evolving, i.e. self-transforming[1] economic system. This fundamental shift of perspective is emphasised by Schumpeter, a pioneer of the study of economic evolution. He tells us that at an early point in his professional life he studied Walras' standard form of economic analysis and discovered that it implied the idea of an economic system which '[i]f it changes at all, it does so under the influence of events which are external to itself'; in other words, 'economic life is essentially passive and merely adapts itself to the natural and social influences' (Schumpeter, 1937/51, 159). This basic approach to economic life was much respected by Schumpeter who praised Walras as the greatest of economists. However, he did not subscribe to the Walrasian perspective. He tells us that he

... felt very strongly that this was wrong, and that there was a source of energy within the economic system which would of itself disrupt any equilibrium that might be attained. ... It is such a theory that I have tried to build and I believe now, as I believed then, that it contributes something to the understanding of the struggles and vicissitudes of the capitalist world and explains a number of phenomena, in particular the business cycle, more satisfactorily than it is possible to explain them by means of either the Walrasian or the Marshallian apparatus. (Schumpeter, 1937/51, 160)

Today we can learn much about the radical change in perspective from Schumpeter. But he would have been the first to admit that there is a long way from an evolutionary perspective and a personally coloured theory to a viable form of evolutionary economics. On the way to what may be called a new evolutionary economics, we need much more than a 'dialogue' with Schumpeter and other representatives of what may be labelled old evolutionary economics.[2] In the present book it will be argued that:

A viable new evolutionary economics is characterised by
(1) a population perspective,
(2) an empirical orientation,
(3) a mix of an algorithmic and a fully formal approach,[3] and
(4) a 'dialogue' with older, verbal studies of economic evolution.

This proposed description of the emerging new evolutionary economics may suitably be explored initially in a specific historical context, namely the period around 1950 which is the year Schumpeter died. Here we have the chance of studying the transmutation of the study of economic evolution as well as of immediately opening the dialogue with Schumpeter. In the latter respect we shall confront one of Schumpeter's last contributions where he presented the empirical orientation of evolutionary economics in a provocative and thought-provoking form which leads to a broader consideration of the relationship between facts and formalisms. In relation to the transition from old to new evolutionary economics we find in the same period an early, non-evolutionary but provocative sketch of an algorithmic approach to economic mechanisms in a paper by Goodwin, which can be used at a starting point for a sketch of this approach. Finally, we also find Alchian's rough and thought-provoking version of the population perspective from 1950.

1.1.1. The empirical orientation as Schumpeter's final thesis

In the history of economic thought and analysis Schumpeter has a paradoxical role: on the one hand, he is probably the sole writer who gave the issue of what we now call economic evolution the pivotal place in his creative analytic work;[4] on the other hand, he is closely connected to the marginalist revolution and to the crowding out of evolutionary perspectives by his agreement to set the standards of economic analysis so high that they cannot be met by evolutionary analysis. Throughout his intellectual career he was tormented by the conflict between his evolutionary pivot and his emphasis on conceptual clarity and mathematical tools of analysis. This is emphasised by him in a couple of letters he wrote in the early 1940s. In the first letter he points out that at an early point in his intellectual career he 'worked out a theory of economic evolution into which personal observation, historical studies, and theoretical work enter in proportions which it is difficult to define.' (Schumpeter, 1991a, 229) In the second letter he states that

> ... there is nothing in my structures that has not a living piece of reality behind it. This is not an advantage in every respect. It makes, for instance, my theories so refractory to mathematical formulations. They can never be so cut and dried as Keynes' schema is; but there are compensating advantages, and one of them is that so many people have told me ...: 'Yes, this is so. I know that from my own experience and observation'. (Schumpeter, 1991a, 230)

However, according to his own criteria of scientific excellence, such an acknowledgement is not enough for a successful theorist. Therefore, Samuelson is right when he points out that

> ... though much of the world regarded Schumpeter as the very essence of an economic theorist, he regarded himself as in a sense a theorist *manqué* [an unsuccessful theorist]. When singing the praises of more exact methods in our beloved science of economics,

Schumpeter claimed he was entitled to do so with a better right since his own work was primarily not in the airy heights of mathematical theory. (Samuelson, 1981, 3)

The reason for this lack of success is mainly due to his chosen area of study rather than to any lack of brightness and analytical skills.[5] Actually, the problems of formalisation of evolutionary arguments and the heavy emphasis on personal and historical investigation is also found in the relevant parts of the works of other economists like Marshall (1890/1961, book 4; 1919) and it is a central characteristic of Darwin's (1859/1964) work on biological evolution.

Such an area did not fit the main trend towards professionalisation of the economics discipline which was strongly supported by Schumpeter: 'One of the many ironies in his life is that his ardent support of mathematics in economics drove his students away from the fields of intellectual endeavor that made his own work so significant, and produced many results that he considered sterile.' (Smithies, 1951, 14) On this background, it is not difficult to understand why the old Schumpeter wanted to present his intellectual testament in historical rather than in mathematical form, even if the latter was his basic ambition (dating back to Schumpeter, 1906/52; 1908).

Less than two months before his death, Schumpeter protested against the crowding out of evolutionary perspectives from economics in a way which shocked not only his young colleagues like Samuelson and Goodwin (Swedberg, 1991, 176) but also broader groups of active researchers in economics and econometrics.[6] The event took place at the Conference on Business Cycles arranged in collaboration between university researchers and the National Bureau of Economic Research (U-NBER, 1951). Schumpeter's point of attack in his paper on the 'Historical Approach to the Analysis of Business Cycles' was implicit but clear to all participants: econometrics and mathematics as the all-dominant tools for the march of modern economics. Schumpeter chose to 'let the murder out'—i.e. to spoil his own performance and create an unpleasant and intellectually troublesome state of affairs—at this conference which gathered a good deal of the most talented young economic theorists, statisticians and model builders of the US:

To let the murder out and to start my final thesis, what is really required is a large collection of industrial and locational monographs all drawn up according to the same plan and giving proper attention on the one hand to the incessant historical change in production and consumption functions and on the other hand to the quality and behavior of the leading personnel. (Schumpeter, 1949/51a, 314)

The purpose of this apparently idiosyncratic proposal of 'detailed historical case studies' (p. 311) was to elucidate the mechanisms underlying much of the cyclical behaviour of economic aggregates. Even the cyclical behaviour of investment is in itself a surface phenomenon and we have to investigate 'the actual industrial process that produces it and in doing so revolutionize existing economic structures.' (p. 312, emphasis removed) The mechanisms of this process had not been studied by Schumpeter's ambitious audience. Actually he thought that

... the most serious shortcoming of modern business-cycle studies is that nobody seems to understand or even to care precisely how industries and individual firms rise and fall and how their rise and fall affects the aggregates and what we call loosely 'general business conditions'. (Schumpeter, 1949/51a, 315)

This lack of interest was especially painful to Schumpeter who was the economist who most clearly had emphasised the evolutionary foundations of the behaviour of economic aggregates. Actually, he had little real intellectual influence even if he had high scores in other respects: 'at the time of his death, a citation index shows that Joseph Schumpeter was the scholar most often cited in the whole field of economics.' (Samuelson, 1981, 1) However, while Keynes' main arguments were theoretically interpreted, developed and applied to a degree where no further reference to the *General Theory* was necessary, Schumpeter got a place in the footnotes. He became

... placed in the category 'footnote economist', i.e. an economist whose works are mentioned in footnotes but are seldom reported and applied more directly in a theoretical interpretation or further development. (Jensen, 1988, 97, my transl.)

Schumpeter's 'final thesis' about the need for industrial and locational case studies was so radically out of touch with the 'logic of the modern scientific situation' (Schumpeter, 1949/51a, 309) that no short-term results could have been obtained. The profession chose a respectful passing over of Schumpeter's 'uncharacteristic performance' (Samuelson, 1951). The reason was that his proposal seemed hopelessly old-fashioned not only because the results of the case studies would be unsuited for mathematical and econometric treatment but also because his emphasis on evolution and leaders looked like romantic-idealistic anachronisms (Swedberg, 1991, 176).

Today we may say that as a representative of the old evolutionary economics Schumpeter was, in a certain sense, an anachronism but his empirical orientation and his emphasis on the heterogeneity principle (which is probably underlying the suggested case studies) is fully in accordance with modern evolutionary studies. His aim appears to have been to influence the trajectory of modelling work by creating an informal demand specification by means of paradigmatic cases and 'stylized facts' (Kaldor, 1961, 178 f.). At least he emphasised (Schumpeter, 1949/51a, 308) that nothing was further from his mind than to start a new Battle of Methods like the one that had earlier raged between marginalist economists and the historical and institutional schools of economics.[7] Instead the task confronting researchers which follows from his advice is ultimately to explore the mechanisms underlying the Schumpeterian type of business cycle, and here no method has a monopoly. 'Theoretical and statistical analysis is in this task as necessary as is historical work. In fact they are inseparable because there is an incessant give and take between them.' (p. 315).[8] The problem for the old

Schumpeter was that balance between the different components had been disturbed and that theoretical and statistical work proceeded without noticing what to him appeared to be the basic explanatory variables.

This balanced approach to theory-evolution seems to be quite modern but this does not make Schumpeter look like a representative of renewed evolutionary studies. Actually, we have to conclude that even if the *Zeitgeist* had not been against him, his proposal to make historical case studies would probably not have helped much as a stand-alone cure against the crowding out of evolutionary perspectives. What was needed was nothing less than a theoretical breakthrough in evolutionary economics which was, in fact, seriously impeded by Schumpeter's emphasis on the historical complexities of evolution and on the explanation of the behaviour of macroeconomic variables. Even Schumpeter's core agent, the innovative entrepreneur who breaks the spell of old routines, was too complicated to start with. Finally, the initial steps of the breakthrough were also impeded by Schumpeter's sharp critique of the use of biological analogies in the exposition of evolutionary-economic ideas (Schumpeter, 1912/34, 57 f.; 1954, 789). For these and other reasons, Schumpeter must rather be considered as belonging to the old tradition of studying economic evolution than to what may be called the new evolutionary economics.[9] However, as soon as a new start is established, Schumpeter's works reappear as a mine of ideas.

1.1.2. The algorithmic approach and Goodwin's prologue

Of special interest is Schumpeter's life-long and fundamental analytical tool problem which has still not left the evolutionary studies in economics and other parts of social science. Even if he provided a 'herculean labor' (Schumpeter, 1954, v), he was not able to overcome this difficulty and live up to his high standards of formalisation. His readers are lucky that he, as mentioned above, developed ways of compensating for the lack of rigour in his theoretical structure. To some extent he transformed some of the difficulties within his chosen area of study into strengths, not least the personal touch and style as well as a richness of historical evidence. But as a theorist he still had a basic 'supply side' problem: the lack of adequate analytical tools for expressing his evolutionary theory compared to the relative abundance of tools suited for other types of studies. In this context the task that Schumpeter and the economists who occasionally tried to help him (Frisch, Samuelson and Goodwin) had set themselves was next to hopeless: Schumpeter insisted on certain characteristics of his evolutionary scheme which could not be supported by the available mathematics or, rather, the available mathematically skilled economists of his time. Goodwin gives an example:

He [Schumpeter] patiently listened to a series of lectures I [Goodwin] gave on Keynesian-type cycle theories, but he would have none of it. Now, half a century later, I better understand why. To begin with they were aggregative, global formulations, which he rightly rejected, since they quite failed to take adequate account of the continuing, innovational re-structuring of the productive economy. Furthermore the whole class of models of that type are unacceptable because they are linear. He perhaps sensed without clearly perceiving this fatal flaw, since his mathematical aptitude was so limited that he barely understood linear dynamic system, let alone nonlinear ones. (Goodwin, 1988, 6)

Schumpeter's lack of mathematical maturity was probably not as much due to a lack of attempts to conquer mathematical tools but rather to his unwillingness to stick to the non-evolutionary assumptions underlying the application of these tools. As Tinbergen (1951/91, 178, 176) has remarked, Schumpeter 'lived another life' and his 'whole attitude vis-à-vis the setting of the problems and their solutions' was radically different from that of the available mathematicians and econometricians.

However, Schumpeter might have had much use for the simple algorithmic approach to the description of economic mechanisms which Goodwin published shortly after Schumpeter's death in the short paper on 'Iteration, Automatic Computers, and Economic Dynamics' (Goodwin, 1951/82). This paper had no immediate relevance for evolutionary economics (and Goodwin has later shown few signs of embracing the population perspective) but he is suggesting that an algorithmic approach to the description of economic mechanisms is an important task in itself. The mechanism in question is the equilibrating mechanism of the Walrasian economic system. But as we shall see, the algorithmic approach is also very helpful by increasing our ability to describe in a clear and flexible way different evolutionary mechanisms. This was exactly what Schumpeter was, in my opinion, searching for from an early point in his life as an economist. Schumpeter's starting point was a methodological analysis of marginalist economics where he emphasised its strictly non-evolutionary nature and suggested a complementary, more-or-less evolutionary research programme (Schumpeter, 1908). On this background, Schumpeter became less interested in Walras' theorems on the characteristics of an economic system in general equilibrium than in an attempt to contrast the mechanisms of innovation with Walras' loosely formulated ideas of the equilibrating mechanism of the economic system. Furthermore, it was of crucial importance to describe the equilibrating mechanism in a simplified manner in order to make possible to insert the complexities implied by rule-changing behaviour into the model. Therefore, the perspectives which were more or less clearly raised in Goodwin's paper suggest an important starting point.

Let us in a modern programming language restate parts of the lecture that Goodwin could have given to Schumpeter. The starting point is Walras' famous words about the equilibrating economic mechanism:

Such is the continuous market, which is perpetually tending towards equilibrium without ever actually attaining it, because the market has no other way of approaching equilibrium except by groping, and, before the goal is reached, it has to renew its efforts and start over again, all the basic data of the problem, e.g. the initial quantities possessed, the utilities of goods and services, the technical coefficients, the excess of income over consumption, the working capital requirements, etc. having changed ... like a lake agitated by the wind, where the water is incessantly seeking its level without ever reaching it. (Walras, 1874/1954, 380)

Goodwin (1951/82, 103) remarks that 'Walras's conception of and term for dynamic adjustment—*tâtonner*, to grope, to feel one's way—is literally the same as that of modern servo theory.'[10] But Walras was not clear as to the character of his picture: sometimes it seems as if it is the economist who gropes for a solution to a system of equations by using the iterative algorithm, sometimes it seems as if we are dealing with an actual economic process where the agents acts in real time. The confusion is serious since Walras is basically dealing with a mathematical method but we may turn it into an interesting question about the actual economic mechanism. This at least is what Goodwin suggests and to some extent explores by means of Wiener's cybernetics study of servo-mechanisms (goal-seeking mechanisms). At that time computer language was not yet invented, but it is not a large jump to try, in the present book, to express the idea in modern terms.

Let us start with a totally naive version of the *tâtonnement* process led by an auctioneer who starts by proposing a trial price vector ($\overline{P}_0$) and then checks whether the quantities of the m commodities supplied and demanded by the n agents would actually clear all markets, if transactions were allowed to take place (i.e., whether $\sum_j D_{ij}(\overline{P}) = 0$, for $j = 1,...,m$). We, therefore, may formulate a 'blind' trial-and-error process in the form of the following algorithm in a semi-formalised programming language:

Naive Tâtonnement:
procedure (agents, i, with commodities, j)
    forbid exchange between i's;
    repeat
        let $P_j := \text{random}(\mathfrak{R}^{+,0})$, for all j;
        invoke $D_{ij}(\overline{P})$, for all i and j;
    until [$\sum_j D_{ij}(\overline{P}) = 0$, for all j];
    allow exchange between i's;
end;

This algorithm will (probably) never stop unless we define a minimum error which is neglected by the economic agents. Even then the probability for finding a solution is incredibly small for a complex economy (see section 4.1.2 on problem complexity theory). There are, however, well-known algorithms (or heuristics in the Artificial Intelligence sense) to make things easier, provided that the excess demand function for each market is well-behaved. We only indicate a simplified version of this algorithm in order to give the flavour of the argument including the Walrasian servo-mechanism, i.e. $\Delta P_j = \sum_j D_{ij}(\overline{P}) > \varepsilon$:

Smarter Tâtonnement:
procedure (agents, i, with commodities, j)
    forbid exchange between i's;
    let $P_j := \text{random}(\Re^{+,0})$, for all j;
    invoke $D_{ij}(\overline{P})$, for all i and j;
    while [$\sum_j D_{ij}(\overline{P}) > \varepsilon$, for some j] do
        let $P_j := P_j - \lambda \sum_j D_{ij}(\overline{P})$, for all j;
        invoke $D_{ij}(\overline{P})$, for all i and j;
    allow exchange between i's;
end;

This algorithm will under the given assumptions converge towards a stop condition consisting of a minimum change of one of the quantities during an iteration. In itself it is not very interesting but the semi-formal approach that it represents makes it relatively easy to formulate a large number of questions for further study: Why is it that the prices and not the quantities (or a mix) equilibrate the system? What happens if the mythical auctioneer is removed and out-of-equilibrium transactions are allowed? By which mechanisms can the agents obtain their apparent foresight? What happens if the behaviour of the agents is relatively inflexible and rule-following? Through experiments with algorithms that try to answer such questions, it becomes patently clear that the Walrasian system represents an extreme case of a non-worldly type. At the same time, it is demonstrated that starting from a confrontation with Walras helps to organise the search for more realistic economic mechanisms and that a programming language may help to articulate the different alternatives. The results of the theories of computability and complexity help to evaluate the abstract possibility of an algorithm as well as its tendency to computational explosion.[11] However, through such studies we gradually drift away from the general equilibrium system and begin to see that the search for a single economic mechanism will never succeed.

So far, we have more or less been following Goodwin. Now we may ask whither this exercise leads. Some might consider this series of studies as a gradual approximation for an evolutionary analysis of economic life. But the underlying heuristic rule is rather peculiar: Instead of guiding theory-evolution by a preliminary version of the searched evolutionary mechanisms by visions, preliminary versions of mechanisms and paradigmatic examples or 'stylised facts', we move forward by looking backward. The research agenda is set by the problems of the Walrasian system rather than by our goal of finding schemes of evolutionary analysis. This appears to be a hopeless strategy for evolutionary economics. Instead we should turn it around by formulating a new goal or 'demand specification' for the mechanisms for which we are searching. Or we may, as the Walras-worshipper Schumpeter might have remarked, look at the next page in Walras' great work where inflexibilities and sudden disturbances of economic life are more important:

It can happen and frequently does happen in the real world, that under some circumstances a selling price will remain for long periods of time above cost of production and continue to rise in spite of increases in output, while under other circumstances, a fall in price, following upon this rise, will suddenly bring the selling price below cost of production and force entrepreneurs to reverse their production policies. For, just as a lake is, at times, stirred to its very depths by a storm, so also the market is sometimes thrown into violent confusion by *crises,* which are sudden and general disturbances of equilibrium. (Walras, 1874/1954, 380 f.)

Analysing this kind of economic process appears to have been one of the driving forces behind the young Schumpeter's (1912) creation of his own theory of economic evolution (*wirtschaftlische Entwicklung*) and his later attempts to substantiate it (Schumpeter, 1939). Here the task is basically to find an evolutionary underlying the shifting periods of prosperity and depression:

... our mind will never be at rest ... until we have assembled in one model causes, mechanisms, and effects, and can show how it works. And in this sense, whatever we may object, the question of causation is the Fundamental Question, although it is neither the only nor the first to be asked. (Schumpeter, 1939, 34)[12]

The importance of the algorithmic approach is that it helps to depict the basic mechanisms which connect cause and effect in an evolutionary process whose concrete path is essentially irreversible. Thus while this approach has basically helped to emphasise the limits of the Walrasian approach, it has a much more positive effect on Schumpeter's research programme. If we may envisage 'Schumpeter with a computer',[13] we recognise that he has obtained a very different expressive power with respect to the 'Fundamental Question' than the real Schumpeter before the computer age. To him new problems will emerge while some of the old problems will simply vanish. Some of these problems will be explored in chapters 3-5 where we shall deal with 'Artificial Economic Evolution'

1.1.3. The population perspective as introduced by Alchian

The transformation from old to new forms of evolutionary economics may be said to have started in the year that Schumpeter died. The programmatic announcement of central themes in new evolutionary economics was made in the paper on 'Uncertainty, Evolution, and Economic Theory' (1950/93) by Alchian, an economist who seems to have been outside Schumpeter's sphere of influence. This now-classic paper—which appears to be based on analogies between biological and economic evolution—presents a revision of neoclassical economics with new assumptions like imperfect information and rule-bound behaviour. Economists have traditionally assumed that economic agents make rational choices or selection among behavioural alternatives. Instead we should 'treat the decisions and criteria dictated by the economic *system* as more important than those made by the individuals in it.' (Alchian, 1950/93, 67) In this new perspective the object of analysis is a heterogeneous set of firms which are subject to a process of selection. In other words, Alchian introduces explicitly into economic analysis what we now call 'population thinking'.

The idea that population thinking is a central characteristic of evolutionary analysis has most clearly been developed in evolutionary biology (see Mayr, 1976, 26 ff.) where this mode of thinking confronts non-evolutionary 'typological thinking'. To think in terms of typologies means to consider the differences between basic types and their concrete instances as something which should be ignored in order to focus on the true essence of the phenomena. For researchers who think in terms of 'populations', things are quite different: the 'typical' characteristics of a population of economic agents are just abstractions while reality is characterised by varying behaviour. More important: this variance is the fuel of the evolutionary process; to ignore it is to retreat to a Platonic world of non-evolving 'ideas'. This retreat is appealing from a purely logical point of view but it makes it impossible to deal with the empirically observable gradual evolution. The same is the case when evolutionary syntheses are crowded out in order to promote logical simplicity. To consider all economic agents as perfectly informed and rational is to consider them as representing a homogeneous type which cannot evolve. Their type is a 'lightning calculator of pleasures and pains' (Veblen, 1898/1961, 73) which responds with varying behaviour to different parameter settings. On the other hand, if their behavioural characteristics (e.g., with respect to degree of profit maximisation) are varying and 'sticky', then there is room for evolutionary processes.

In Alchian's population perspective we consider a large number of different agents and, therefore, we tend to abstract from the processes of individual decision-making or at least to consider them in a highly simplified form. In the most simplified form, we have a Darwinian-like trial-and-error process. Alchian illustrates some parts of such a process through the example of the behaviour of a population of travellers:

Assume that thousands of travellers set out from Chicago, selecting their roads completely at random and without foresight. Only our 'economist' knows that on but one road are there any gasoline stations. He can state categorically that travellers will *continue* to travel only on that road; those on other roads will soon run out of gas. Even though each one selected his route at random, we might have called those travellers who were so fortunate as to have picked the right road wise, efficient, foresighted, etc. (Alchian, 1950/93, 69)

This set-up is made to demonstrate that totally random behaviour may be filtered by a selection mechanism in a way which makes the observable behaviour look as if it is informed and well-thought-out. Our omniscient economic researcher is shocked by the many attempts to conclude on the appearances and to rationalise the behaviour of the travellers. But he is able to explain and, if he is informed, to predict changes in the behaviour:

> If gasoline supplies were now moved to a new road, some formerly luckless travellers again would be able to move; and a new pattern of travel would be observed, although none of the travellers had changed his particular path. The really possible paths have changed with the changing environment. All that is needed is a set of varied, risk-taking (adoptable) travellers. (Alchian, 1950/93, 68 f.)

Even these changes may be misinterpreted: the correspondence between changes in behaviour and the economic environment appears to suggest that the travellers are able to learn and keep their acquired characteristics. But, according to Alchian, this is really a case of 'Lysenkoism', i.e. mistaken Lamarckism where a Darwinian account would have been correct. On the other hand, it should be remarked that Alchian's emphasis on a random mechanism in the process of problem solving at the population level is extremely wasteful: most of the travellers have to walk back to Chicago but some solve the problem of getting away from the city. It is also very demanding with respect to the irrationality of the agents: if all travellers are learning not to use the non-supplied roads by doing the first experiment, then nobody will explore the new road where gasoline is present in the second experiment. This version of the system is characterised by a lock-in situation where no long-distance travelling takes place. The presence of an absorbing state for the system demonstrates that we are not facing a full analysis of an open-ended evolutionary process and this may be the reason why Alchian did not explore a Lamarckian mechanism of learning by doing and the transmission of acquired characteristics.

When Alchian deals with economic agents rather than imaginary travellers it is important to ask for the 'criterion by which the economic system selects survivors' (Alchian, 1950/93, 67). The obvious answer is that the criterion is the level of realised profits of the individual firms: 'those who realize *positive profits* are the survivors; those who suffer losses disappear.' (*ibid.*) Like in the case of Darwinian natural selection, we should not presuppose that the success or survival of individual firms are based on full information about the rules of the game and about the other agents in it. The question of success is decided *ex post* rather than *ex ante*: 'Among all competitors, those whose particular conditions happen to be the most appropriate of those offered to the economic system for testing and adoption will be "selected" as survivors.' (pp. 67 f.) Thus, by apparently minor modifications of the standard economic account Alchian arrives at 'a vastly different analytical framework—one which is closely akin to the theory of biological evolution.' (p. 74) Within this framework we can give

> ... accounts for observed uniformity among survivors, derived from an evolutionary, adopting, competitive system employing a criterion of survival, which can operate independently of individual motivations. (Alchian 1950/93, 73)

The framework for developing such explanations is in principle independent of the biological analogy (Alchian, 1953) but it is by means of this analogy that it gets its appeal (Friedman, 1953) as well as its adversaries (Penrose, 1952). By suggesting that the 'economic counterparts of genetic heredity, mutations, and natural selection are imitation, innovation, and positive profits' (Alchian, 1950/93, 74), whole trains of thought are started in interested readers. But the really interested reader soon runs into trouble which primarily relates to the fact that there is no real analogy between 'genetic heredity' and 'imitation'. The introduction of the possibility of 'imitation' makes the firms much more adaptable than they would be in a Darwinian process of 'survival of the fittest'. Actually, the firms may innovate and imitate more rapidly than the competitive process is selecting the 'fittest' and in this case the outcome will not be an evolutionary process but a state of behavioural chaos. They may also be so slow in changing relative to the functioning of the selection process that the system becomes characterised by a non-evolutionary stasis (only slightly disturbed by exogenous change). Unfortunately, Alchian does not demonstrate how long-term evolutionary processes may take place between the Scylla of behavioural chaos and the Charybdis of behavioural stasis. For this reason his programmatic announcement of an evolutionary renewal is radically incomplete and it might be reabsorbed in non-evolutionary modes of thinking.

These limitations of Alchian's approach suggest the premature character of Friedman's early attempt to accommodate neoclassical economics to Alchian's ideas (or *vice versa*) in his famous paper on 'The Methodology of Positive Economics' (1953). Here Friedman simply presupposes that the process of selection is strong and fast enough to remove the non-adapted types of behaviour:

> Let the apparent immediate determinant of business behavior be anything at all—habitual reaction, random chance, or whatnot. Whenever this determinant happens to lead to behavior consistent with the rational and informed maximization of returns, the business will prosper and acquire resources with which to expand; whenever it does not, the business will tend to lose resources and can be kept in existence only by the addition of resources from outside. The process of 'natural selection' thus helps to validate the hypothesis [of profit maximization]—or, rather, given natural selection, acceptance of the hypothesis can be based largely on the judgement that it summarises appropriately the conditions for survival. (Friedman, 1953, 22)

This argument, which has convinced generations of economists, reflects that the verbal account of an evolutionary process easily leads to a neglect of central assumptions. Winter (1964, section II; 1971/93, section III; 1975, 96-99) has provided a list of some of the implicit assumptions beneath Friedman's argument which lead to serious questions—at least if one has Winter's realistic view of scientific explanation: Are we facing 'habitual reaction' which may allow the process of selection to take place or 'random chance, or whatnot' where we cannot assume that a successful mode of behaviour will be upheld? Can we assume that the firms with the right (profit-maximising) behaviour are the ones which actually expand? Is the process of variety-creation sufficiently strong to exclude long-term persistence of suboptimal modes of behaviour? How fast is the process of selection of the optimal behaviour?

To Friedman's instrumentalist view of scientific models, such questions do not appear to be particularly relevant since theories are not expected to be realistic, as long as they produce correct predictions about the future. However, from the viewpoint of the development of a new evolutionary economics they are quite central. Actually, they led Winter to the creation of much more satisfactory outlines of evolutionary economics based on the population perspective than Alchian had delivered. In these outlines (Winter, 1964; 1971/93) we see the beginnings of a synthesis between Alchian's study of the selection mechanism, Simon's work on behavioural routines and Schumpeter's ideas on innovative behaviour; this synthesis will be explored in chapter 4. But even today the problems raised by Alchian's introduction of population thinking into the realm of economic analysis have not been fully explored.

1.2. Surveying evolutionary economics[14]

1.2.1. Crude definitions

The difficulties of evolutionary-economic theories are, not least, due to their attempt to account for the *endogenous* transformation of the knowledge applied in economic systems. More specifically, the problems spring from the attempted explanation of the ever-changing diversity and adaptedness of the decision rules, commodities, production methods, and organisational forms of economic life. The endogenous explanation of these facts (the 'data' of standard neoclassical analysis) will necessarily focus on the mechanism of the evolutionary process: given an outcome of a supposedly evolutionary process, we try to point out the mechanism which has brought about this outcome or result. The necessity of this mechanism-oriented approach to explanations (Elster, 1983; 1989) is based on the fact that novelty plays an essential role in evolutionary processes. Shackle has emphasised that this characteristic of evolution excludes standard modes of scientific explanation:

Whether evolution, irreversible, non-repetitive change, can in logic lend itself to analysis is open to doubt. Science depends upon regularities, upon repetitions of path, structure or association, and evolution in its long-range effects is the negation of repetitive phenomena. If there can be a theory of evolution of any subject-matter, whether biological or social, it must lie in mechanism, in the basic principle which allows or directs an explicatory development. (Shackle, 1965, 187 f.)

From this viewpoint the important thing is to develop the tools for a realistic study of the mechanism of economic evolution and at the same time to shift away from the instrumentalist methodology of economics with its concentration on prediction of future outcomes and its neglect of realistic mechanisms (propagated by Friedman, 1953).

Evolutionary-economic studies have a synthetic character largely because the mechanism underlying economic evolution is very complex; this mechanism should be thought of as a synthesis between different (sub)mechanisms rather than as a single mechanism whose parts can be considered as black-boxes. A basic task is to show how an evolutionary process can be synthesised from these individual mechanisms. The major difficulty is due to the fact that the mechanisms are normally considered to be related to the domains of different scientific disciplines: the preservation and transmission of rules and norms have traditionally been related to sociology but are also studied by, e.g., institutional economics; the mechanisms of variety-creation are studied by psychology and social psychology, but also as interdisciplinary innovation studies which include a heavy economic component; the mechanisms of selection have especially been studied by standard economics, both as the competition between economic agents and as the agent's choice between different alternatives; the mechanism of segregation or closure is an area of industrial economic dynamics as well as of sociology, etc. The attempted integration of these diverse mechanisms in the study of evolutionary processes clearly represents an ambitious and risky synthesis. Representatives of the different specialities tend to emphasise 'their' mechanism at the cost of other mechanisms. But it is the synthesis between different theories rather than the contributions to the detailed understanding of the individual mechanisms which is the core factor of evolutionary economics. This combination of different mechanisms may even be taken as a definition:

> *An evolutionary-economic explanation* is an explanation of a fact of economic life by reference to previous facts as well as to a causal link which (immediately or in reconstructed form) may be shown to include
> (1) a mechanism of preservation and transmission,
> (2) a mechanism of variety-creation,
> (3) a mechanism of selection,
> and which includes or may be enhanced by introducing
> (4) a mechanism of segregation between different 'populations'.

The emergence of an evolutionary process presupposes that none of the individual mechanisms becomes too dominant. If preservation dominates, the result is a stasis of economic knowledge, while a dominance of variety-creation leads to non-deterministic chaos. We need certain assumptions about the economic agents in order to generate a system of such agents who can change their behaviour in an irreversible manner through a self-generated process. Such an assumption-oriented delineation of evolutionary economics tends to be more restrictive than the mechanism-oriented definition and we can only hope to reflect a limited part of the studies within the realm of evolutionary economics. Still, it is useful to produce a loose list of:[15]

Typical assumptions and characteristics of evolutionary-economic explanations

1. The agents (individuals and organisations) can never be 'perfectly informed' and they have (at best) to optimise locally rather than globally.
2. The decision-making of agents is normally bound to rules, norms and institutions.
3. Agents are to some extent able to imitate the rules of other agents, to learn for themselves and to create novelty.
4. The processes of imitation and innovation are characterised by significant degrees of cumulativeness and path dependency but they may be interrupted by occasional discontinuities.
5. The interactions between the agents are typically made in disequilibrium situations and the result is successes and failures of commodity variants and method variants as well as of agents.
6. The processes of change occurring in a context described by the above assumptions and characteristics are non-deterministic, open-ended and irreversible.

This list may look as if it has been constructed to contradict the standard neoclassical assumptions. At present it should, however, be seen as yet another indicator of the difficulties of evolutionary economics: it is extremely difficult to deduce mathematical results from such a basis. Instead the list emphasises the empirical orientation of evolutionary economics. But this empiricism cannot be upheld in all cases and the theoretical development of evolutionary economics will often presuppose much stricter and less realistic assumptions.

This list of assumptions and characteristics also raises another serious question: to what extent is it helpful to search for *typical* evolutionary explanations? If we apply population thinking to the subject of the development (or evolution) of evolutionary-economic theory itself, it may—at least in certain periods—be more helpful to emphasise the variety or diversity of evolutionary explanations. Even if the ultimate aim is to create a modern evolutionary-economic synthesis, any premature lock-in of the theoretical development may be very harmful. The quest for crude definitions is, however, unavoidable. In particular, there is a natural tendency to neglect the internal differences within evolutionary economics in order to make it easily comparable to dominant modes of economic analysis, especially neoclassical economics.

This tendency is clearly found in the presentation of the characteristics of evolutionary-economic explanation in contradistinction to neoclassical-economic explanation in Nelson and Winter (1982, chs. 1, 2, 3, 6, 8, 15 and *passim*). Their strategy of exposition helps to bring out some of the programmatic claims of evolutionary economics and it faces squarely the fact that the basic scientific training of most of the students and researchers in the area is based on neoclassical modes of explanation. However, the complexities of the comparison between evolutionary and neoclassical economics presuppose simplifications with respect to the two entities which are being compared. The most obvious simplification is the concentration on 'textbook neoclassicism' rather than 'research-paper neoclassicism' (see Winter, 1991) but more serious simplifications are based on the tendency to avoid the differences and difficulties within evolutionary economics in order to ease the comparison with neoclassical explanation.

1.2.2. On the modern history of evolutionary economics

The crowding out of evolutionary perspectives (which is reflected in Schumpeter's 'final thesis') is an interesting but little studied aspect of the history of economic thought and analysis. This crowding out of 'the Mecca of the economist' was certainly not intended by Marshall (1898, 43),[16] as little as it was by Adam Smith, Marx, Menger, or Schumpeter.[17] On the contrary, they saw their work as first approximations to a fuller understanding of the evolution of economic life and they gave very diverse suggestions of how to approach economic evolution: Smith (1776/1922) gave hints about the interplay between division of labour, dynamic economies of scale and accumulation of capital (see Young, 1928; etc.); Menger (1871/1981; 1883/1985) insisted on dealing with difficult subjects like the changing quality and diversity of economic goods as well as the non-designed emergence of institutions like money through a process of trial-and-error (O'Driscoll, 1986); Marshall (1890/1961, book IV and appendix H) emphasised external and internal economies of the 'representative firm' in his long-term analysis of supply, which should later be enriched by an explicit analysis of the rise and fall of unequally endowed firms in their struggle for life (Thomas, 1991); Schumpeter (1912/34; 1939; 1942/87, part II) presented a dualistic vision and analysis of economic evolution as a struggle between on the one hand innovative entrepreneurs and bankers and on the other hand rule-based economic agents.

But the result of these uncoordinated suggestions was not the emergence of a more integrated approach. Instead the result was exactly opposite of their intentions, as if the economic profession was led by an 'invisible hand'. Many of the attempts to eliminate the paradoxes, find the core arguments, clarify and axiomatise had as the first victims the hints about the evolutionary process, like increasing returns, representative firms, industries and commodity types, profit-creating entrepreneurship, etc.[18] In all cases the evolutionary attempts could be seen by main-stream economists as representing sources of confusion, as misfit parts of economic analysis; in the language of fairy tales they were the ugly ducklings in the yard of economics. The marginalist revolution, the Keynesian revolution and the post-war formalist revolution each have made their contribution to the crowding out of evolutionary-economic offspring. As a result, the multiform old evolutionary economics had largely disappeared in the first decades of the post-war period. Even the notion of 'evolutionary economics' which was first used by Veblen in his famous article 'Why is Economics Not an Evolutionary Science?' (1898/1961, 76 f., 79) was almost forgotten, except as a seldom used synonym for institutional economics in the tradition of Veblen and Commons.

The failure of many different openings toward the study of economic evolution cannot be blamed solely on external factors like the dominance of the neoclassical paradigm in the economics profession.[19] This kind of explanation of the crowding out tends to draw the attention away from fundamental difficulties which are intrinsic to such a study. Instead we should emphasise that such difficulties are probably sufficient for any scientific observer to cast serious doubts about the feasibility of an evolutionary-economic research programme. It will suffice to mention a few problems: the very limited predictability of the outcome of evolutionary processes which appears to block the falsification of evolutionary theories (Popper, 1972); the synthetic or impure character of the evolutionary mechanism which forces evolutionary-economic theories to transgress the borders of different social-science disciplines and creates an impression of eclecticism; the historical or empirical orientation of evolutionary explanation which hinders any clear-cut demarcation line toward business history and the do-it-yourself economics of practical economic life. Such difficulties are probably sufficient to explain the crowding out of evolutionary-economic perspectives. They relate to a mode of thinking which is difficult to learn, suggests messy 'dialectical' concepts (Georgescu-Roegen, 1971), blocks convenient modelling strategies, and blurs border-lines to other disciplines. Had it not been for the huge potential explanatory power of a successful theory of economic evolution, the difficulties would probably have been sufficient to scare away all researchers from the evolutionary field.

Today we see clear signs that the 'dark ages'[20] for the evolutionary perspectives are over. One sign is the still more frequent citing of the different types of old evolutionary economics (relating to Menger, Marshall, Veblen, Schumpeter, and others). Another sign is the rapid spread of economic studies emphasising the structural similarities between biological and economic analysis with a special emphasis on evolutionary issues (surveyed in Khalil, 1992; see also Rosser, 1992). A still more important sign is the emergence of studies of economic and related technological and institutional phenomena which are based on an explicit and clear concept of an evolutionary process, a concept which was not really present in the old evolutionary economics. Such studies may be classified as instances of what I shall call the emerging new evolutionary economics. In other words, the new evolutionary economics is here defined in a preliminary way by its application of an explicit and clear concept of an evolutionary process.[21]

The first applications in economics of a well-articulated idea of an evolutionary process are found in Alchian's (1950/93) and Winter's (1964; 1971/93) programmatic papers (see section 1.1.3), but the real change with its focus on evolutionary mechanisms has taken place in the last 10-15 years. A landmark in this respect was made when Nelson and Winter rewrote several earlier articles and formulated a research programme in their book on *An Evolutionary Theory of Economic Change* (1982).[22] This work showed the possibility and importance of a new start, although not all agreed with their particular strategy based on the assumption that the 'verbal account of economic evolution seems to translate naturally into a description of a Markov process—though one in a rather complicated state space.' (p. 19) But this way of formulating their synthesis made easy the next step, namely the translation of the Markov process to computer models and simulations which allowed a treatment of intricacies of evolutionary mechanisms not imagined by, e.g., Alchian. In this way Nelson and Winter succeeded in formulating a family of models where firms are not only 'naturally selected' by the economic system; the firms are also influencing their own destiny by modifying their own behaviour through processes of search for and selection of new modes of behaviour. Thus Nelson and Winter provided constructive proofs of the existence of relatively interesting evolutionary-economic models. However, very few researchers have followed Nelson and Winter's work directly. (See chapter 4)

Another beginning which is more important in relation to general economics is related to new developments in relation to game theory. Models which appear to be well-suited for the study of evolutionary processes have been created, e.g. in the form of repeated games of players with limited information and calculating capabilities, with procedural behaviour and learning from past experience, etc. (see Aumann, 1987, 468 f., 478; Kreps, 1990a, ch. 6) In this area there has been an interaction between economic and biological applications, not least relating to the concept of an evolutionary stable strategy, i.e. a strategy which when dominant in a population cannot be outcompeted by an alternative strategy (Smith, 1982). We now have an important literature on the properties of evolutionary games in economics (Sugden, 1989/93; Friedman, 1991). However, one of the most innovative examples of how to apply iterated games in both theoretical and experimental ways is found in Hirschleifer (1982/93; Hirschleifer and

Coll, 1988; 1992) and especially in Axelrod's work on *The Evolution of Cooperation* (1984; Axelrod and Dion, 1988). In the latter case the different strategies are tested in an iterated Prisoner's Dilemma while they are generated either by different game theorists (who participate in computer tournaments) or by the computer itself by means of so-called genetic algorithms. (See chapter 5)

An important issue in the analysis of evolutionary-economic processes is how history matters (path dependency). That this is the case appears obvious for any consideration of the facts of economic evolution. But it is not easy to treat this intuition systematically. Attempts to do so have especially been made by Arthur (1988; 1993; Arthur *et al.*, 1987/93) and David (1985/93; 1993). They demonstrate by means of stochastic processes how the development of a system with an initial variety of possibilities can show a lock-in which blocks further adaptation. The empirical standard example is the process which led to the emergence of the QWERTY keyboard even if for decades it has been known to be a suboptimal solution to the typing problem. More generally, the evolutionary process is seen as taking place in a 'landscape' where it is bound to find locally rather than globally optimal solutions. This result seems to substantiate the idea of an evolutionary process which normally follows the trajectories created by the constraints of given paradigms but which occasionally shows more basic alternatives (see sections 2.5-2.7).

Another issue is how to treat the extremely complex coevolution of different commodities, production methods or institutions. Here it is necessary to have a more distant approach to evolutionary processes than in the case where every economic decision is treated explicitly. However, in the new evolutionary economics there has been a tendency to have too much evolution and too little 'ecology', too much variety and too little structure. Instead the interaction has often been left to non-evolutionary economists. The first step to change this situation has been to reconsider the huge literature on the diffusion of innovations within an evolutionary framework (Metcalfe, 1988). The next step is to deal with several interacting populations which are partly competing, partly oriented towards different static or dynamically changing 'ecological niches'. Here the work has hitherto been dominated by business economists and organisation theorists (like Hannan and Freeman, 1989). The problem at this level of analysis is not only the complexity of the interactions but also the lack of economic theories on commodity types and other 'mesoeconomic' concepts. Given such a theory, there is a possibility of transforming the results of modern evolutionary-ecological biology in a way which helps to systematise older evolutionary-economic ideas of growth poles and development blocks (Dahmén, 1950/70; 1988) and recent ideas of national systems of innovation (McKelvey, 1991; Lundvall, 1992; Nelson, 1993). (See chapter 3)

Many other studies of relevance to the emergence of new evolutionary economics could be cited. First, we find studies emphasising structural similarities between biological and economic analysis. Such studies help to transfer to economics the know-how of coping with the general methodological difficulties of evolutionary studies. Especially, they help to explore the surprisingly difficult population perspective (and the related heterogeneity principle) which underlies much of evolutionary economics. Second, there exists a huge empirically oriented literature on the economics of technological and institutional change, which is more or less clearly related to the new evolutionary economics. Basically, evolutionary economics is an empirical science which deals with the intellectual reconstruction of the real mechanisms of economic evolution; its models must be explored and developed in interaction with well-defined areas of empirical investigation. Unfortunately, the work on *Technical Change and Economic Theory* (Dosi *et al.*, 1988) is unique in the way in which it brings together empirical and 'grounded theoretical' work with the more abstract attempts to develop new evolutionary economics. Third, there is the rapidly expanding literature on Menger, Hayek, Schumpeter, and others. Especially, the concept of an evolutionary process is used to give a new account for the history of evolutionary-economic theory in retrospect (to paraphrase the expression of Blaug, 1962/85). All these studies are important. However, the emphasis of the present progress report on the conceptual and formal tools reflects a central tendency in the literatures as well as a central thesis of the present book.

Seen from a modern point of view the major problem with the old evolutionary synthesis is that it is built of out-dated elements and that it is so complex that it is nearly impossible to master intellectually. We see, e.g., that 'the intellectual coherence and power of thinking about Schumpeterian competition have been quite low, as one would expect in the absence of a well-defined theoretical structure to guide and connect research.' (Nelson and Winter, 1982, 29) The research agenda for the new evolutionary economics has been dominated by a search for such a 'well-defined theoretical structure' and thus by a wish to overcome the short-comings of the old evolutionary economics. The introduction of such systematic reconstructions of evolutionary processes helps to bring the latent ideas of the old evolutionary economics into clear consciousness and thus to overcome some of its intrinsic problems. The relatively approach is, so to say, acting as a midwife for the birth of an evolutionary paradigm in economics (and other social sciences)—especially by facilitating the clear articulation of the surprisingly difficult concept of an evolutionary process.

This thesis about the central role of a relatively formal approach to the study of economic evolution is apparently diametrically opposed to the lesson from the history of the crowding out of evolutionary perspectives; experience suggests that a further formalisation may tend to close the niches for alternative thinking. At best one might suspect that the resulting new evolutionary economics will be a highly distorted version of the pre-paradigmatic ideas of the old evolutionary economics. The result would be yet another form of crowding out, this time of the 'magnificent dynamics' (Baumol, 1951/70, 8 f., 21) of economic evolution as studied by Schumpeter and others; they would be substituted by limited and boring forms of evolutionary analysis. The present book demonstrates that there are good reasons for this expectation but it also shows that the necessary but painful transformation to a paradigm-based new evolutionary economics may be eased by tools which mediate between, on the one hand, the informal and empirical approaches and on the other hand the fully mathematicised analysis of evolutionary processes. Actually, some of the studies which have been cited represent such a mediation rather than a fully formal approach. They are more or less representative of a flexible algorithmic approach to evolutionary economics.

However, even in the presence of much theoretical flexibility, it is a major problem that the relative success of the new evolutionary economics has, until now, been obtained by redefining the goals of evolutionary economics in a way which tends to increase its isolation from historians and policy makers (David, 1993). In this sense, much of the evolutionary-economic work has some similarities with the mathematical modelling of biological evolution in the 1930s, which was only understood by very few researchers but nevertheless became a major element of modern evolutionary-biological explanation. In this comparison we are still waiting for something similar to the 'modern evolutionary synthesis' between practically oriented naturalists and theoretically-experimentally oriented geneticists (as described in Mayr and Provine, 1980). It is one of my central theses that the recent evolutionary-economic modelling may be developed into a part of a similar overall, modern synthesis between descriptive and theoretical studies of the diversity and adaptiveness of economic life. The algorithmic rethinking of major contributions to the new evolutionary economics should be seen as stepping stones on the way to such a modern synthesis. In this perspective is important that the 'dialogue' between new and old evolutionary economics is not necessarily bound to take place at a theoretical level. The old evolutionary economics was to a large extent an empirically oriented endeavour (see, e.g., Marshall, 1919; Schumpeter, 1939). Thus the 'dialogue' may also help to create a balance and a fruitful but difficult interaction between theoretical and historically oriented work and thus to promote a modern evolutionary-economic synthesis.

Notes

1   The idea of a self-transforming system is emphasised by Schumpeter. A recent statements of the importance of this idea is found in Witt (e.g. 1991a; 1993b).
2   The division of the studies of an area of science into an 'old' and a 'new' part is a well-known and not always fruitful means of exposition. The 'new' is often thought of as the neoclassical, like in new economic history. The present usage is more

related to the discussions about old and new institutional economics, see Hodgson (1989) and Langlois (1989). However, I emphasise differences with respect to formalisation and modelling which are not found in Hodgson's distinction between on the one hand the 'old' tradition of Veblen and most of American institutionalism and the extremely heterogeneous new institutional economics of Williamson, Schotter, Hayek and others. In my interpretation, most of the contributions to neo-Austrian evolutionary analysis (including Hayek) represent an old evolutionary economics.

3   The present book concentrates on the algorithmic approach to evolutionary economics and thus on an important distinguishing mark. This emphasis is not meant to deny the importance of a fully formal approach or the interplay between algorithmic and formal formulations.

4   The present interpretation of Schumpeter's relationship to evolutionary theory in a modern sense is not accepted by all researchers. An alternative view is, e.g., put forward by Hodgson (1993, 149) who suggests that 'the invocation of Schumpeter's name by the new wave of evolutionary theorists in the 1980s and 1990s is both misleading and mistaken.' In my opinion this statement is built on too simple an understanding of Schumpeter's relationship to Walras (Hodgson, 1993, 140 ff.) as well as on an overestimation of certain aspects of Schumpeter's theory which do not fit into a Darwinian scheme of evolution (pp. 145 ff.). At present, it should just be noted that when dealing with the usage of the notions 'evolutionary' and 'developmental' in social science, scholars are facing a terminological quagmire which I try to avoid in different ways in the following sections and chapters. Here I can only try to show that, late in his life, Schumpeter was not against being characterised as an evolutionary theorist. Luckily, he has even stated a criterion for 'evolutionism' which clearly covers what we find in his major works like *The Theory of Economic Development* (1912/34) and *Business Cycles* (1939). According to this criterion, evolutionism means far more than the proposition: 'Social phenomena constitute a unique process in historic time, and incessant and irreversible change is their most obvious characteristic.' (Schumpeter, 1954, 435) Such a statement was to Schumpeter simply a recording of facts. To be an evolutionist researcher means to go further and make evolution 'the pivot of one's thought and guiding principle of one's method.' (p. 436) I will argue that Schumpeter was such a researcher whose thought '*turned* upon evolution' (p. 436), who tended to see evolution as the pivotal question on which everything turns. That Schumpeter also thought of himself in this way, we see in short statement of his major scientific contribution which he wrote in a letter some months before his death: '... I began at an early age to look upon economic life essentially as a process of change, and I tried to make the main features of this change the centre of my own type of theory. In doing so, I discovered that a number of phenomena such as entrepreneurial profits, interest, and credit found ready explanation within such an evolutionary schema. The theory of business cycles had, of course, always been "evolutionary" by nature, but even in this case a new explanation occurred to me that differed from others in showing that the mechanism of evolution so works as to produce a wave-like movement of its own even when there are no external disturbances to produce them.' (Schumpeter, 1991a, 235) He was, of course, thinking of a special kind of evolutionary vision and theory. It is not the wide-spread theory of gradual evolution but a theory which underlines that evolution is 'more like a series of explosions than a gentle, though incessant, transformation.' (Schumpeter, 1939, 102) See discussion in chapter 2.

5   The idea of Schumpeter's dilemma between his evolutionary ideas and a paradigm of 'good science' is, as far as I can see, not developed in any of the increasing number of studies on Schumpeter and his work, including the following monographs: Perroux, (1935/65); Clemence and Doody (1950/66); Marty (1955); Khan (1957); Schneider (1970); Oakley (1990); Allen(1991); Swedberg (1991); Bottomore (1992).

6   The shock could have been avoided, if other economists had followed Schumpeter's work in the late 1930s and in the 1940s more closely than they did. *Business Cycles*

(1939) clearly emphasises the historical method, and in his last years Schumpeter got involved with the new Research Center in Entrepreneurial Studies which suggested new developments of his ideas (Swedberg, 1991, 172 ff.). A possibility of the revitalisation of his theoretical scheme through the interaction with theoretically relevant studies in industrial history appeared to emerge.

7   Schumpeter has written much about the famous Continental *Methodenstreit,* but when facing a US audience he was primarily referring to the US controversies between abstract economic theory and the empirically oriented Institutionalism of especially Mitchell, the founder of the National Bureau of Economic Research, who should have led the conference but died before it was held.

8   This is a shortened version of the more developed methodological viewpoint in Schumpeter (1954, ch. 2),

9   Later in the book it will become clear that both the old and the new evolutionary economics are characterised by a considerable degree of heterogeneity/variety (see, e.g., chapter 6).

10 A servo-mechanism is an error control mechanism like a thermostat. It acts upon differences between the desired and the actual value of some control variable in a way which diminish the error.

11 To experiment with the complexity of the algorithms we may introduce a 'time counter' in the repeat-loops and the while-loops of the algorithms, indicating one time interval for each iteration. But this aspect of time is not allowed to have any real consequences in the Walrasian framework since all exchanges are forbidden during the iterations. He is clearly working in a timeless system or in pseudo-time.

12 Here we see Schumpeter as much of a methodological realist. This view may be considered as a self-criticism *vis-à-vis* his instrumentalist stand in Schumpeter (1908) where he tries to persuade us to discard the concept of causality from economic theory. This change does not appear to have been noticed in Shionoya's (1990) paper on Schumpeter's instrumentalism but it is mentioned by Machlup (1951/91, 235). The realist focus on causal mechanisms as an essential part of explanatory work is clearly stated the philosopher Elster (1989, 3): 'To explain an event is to give an account of why it happened. Usually, and always ultimately, this takes the form of citing an earlier event as the cause of the event we want to explain, together with some account of the causal mechanism connecting the two events.'

13 A similar picture was developed in the heated debates over *The Limits to Growth* models (Forrester, 1971; Meadows *et al.*, 1972). Here it was said that we were simply facing 'Malthus with a Computer' or rather, a case of 'Malthus in, Malthus out' (Freeman, 1973, 8). However, it was not emphasised that a repeated use of the computer leads to a transformation of the Malthusian ideas rather than a simple resurgence of them.

14 Compare the broader surveys by Witt (1991a; 1991b). The present survey emphasises the definition and historical development of evolutionary economics.

15 Compare, e.g., Clark and Juma (1987), Saviotti and Metcalfe (1991), and Dosi (1991).

16 The full version of Marshall's famous statement is: 'The Mecca of the economist is in economic biology rather than economic dynamics.' Marshall is not primarily thinking of evolutionary theory in a Darwinian sense but rather of Spencer's ideas of evolution which not only include the 'struggle for life' but also a progress towards complex forms of organisation, see Moos (1990).

17 The different evolutionary perspectives are presented in recent reviews of the history of evolutionary thought in economics, see Clark and Juma (1988) and Hodgson (1993). The crowding out of such perspectives is also hinted at in these works. It is also dealt with more or less explicitly in, e.g., Veblen (1899/1961), Schumpeter (1908), and Shackle (1967).

18 All these suggestions have their clear limitations. As an example we may take the notion of a 'representative firm' which, in relation to Marshall's story of the typical tree in the ever-changing forest, reminds us of the evolutionary process underlying the development of the industry. However, for a further development of evolutionary

analysis this notion seems to be a hindrance since it draws attention away from the heterogeneity which is the fuel of evolution.
19 This appears to be the implicit assumption behind the argument in, e.g., Nelson and Winter, 1982; Clark and Juma, 1987; some of the papers in Dosi *et al.*, 1988.
20 Similar but shorter periods have been found in other social sciences. Sanderson (1990, 2) has summarised the upswing and downswing in evolutionary thinking within the social sciences in the following way: 'The heyday of evolutionism was in the second half of the nineteenth century, for it was then that the doctrines of Morgan, Tylor, Spencer, Marx, and others were produced. This "golden age" of evolutionary social science came rather suddenly to an end shortly after the turn of the century, however, and the first decades of the twentieth century represented a sort of "dark age" for evolutionism. During this time evolutionism was severely criticized and came to be regarded as an outmoded approach that self-respecting scholars should no longer take seriously. Evolutionary theories did not die out completely, but they were seldom seen, and even the word "evolution" came to be uttered at serious risk to one's intellectual reputation. Antievolutionism, rather than evolutionism, was the watchword of the day. [/] Yet the reign of antievolutionism was itself to last no longer than had the evolutionism that preceded it. By the 1930s some scholars were beginning to take evolutionism seriously again, and by the 1940s an "evolutionary revival" was well under way.' Sanderson's summary of the destiny of evolutionary thinking explains the cautiousness of many economists working with evolutionary problems in the first half of this century. For example, we see that Schumpeter remarks that 'the evolutionary idea is now discredited in our field' and that he does not endorse 'unscientific and extra-scientific mysticism' and 'dilettantism'; therefore, 'we must be careful with the phenomenon [of evolution, *dem Entwicklungsphänomen*] itself, still more with the concept in which we comprehend it, and most of all with the word by which we designate the concept and whose associations might lead us astray in all manner of undesirable directions.' (Schumpeter, 1912/34, 57 f. [supplemented by Schumpeter, 1912/26, 88])
21 While frequently used, the notion of evolutionary economics is still lacking a widely agreed upon definition and we are still waiting for an authoritative exposition of 'the elements of evolutionary economics'. In this situation the inclusion of the term evolutionary economics into the title of a book or paper gives little information of its contents. In one case (Tool, 1988) the name reveals itself as a seldomly used synonym for American institutionalist economics; in another book from the same year the name is indicating neo-Schumpeterian economics (Hanusch, 1988a). We also find attempts to combine post-Keynesianism and institutionalism into 'evolutionary macroeconomics' (Foster, 1987) and to use evolutionary economics as a label for Boulding's (1981) application within economics of a wide range of ideas from evolutionary biology and systems theory. But in what is probably the most well-known work in the area, the book by Nelson and Winter (1982) we find only more modest formulations like an 'approach' or 'an evolutionary theory of economic change', while the name 'evolutionary economics' is first met in the index of the book (p. 432) where it suddenly becomes a designation of the Nelson and Winter research programme which combines ideas of Alchian, Simon and Schumpeter and a number of novelties. Other kinds of 'evolutionary approaches' are represented by authors who see sociobiology and iterated games as important sources of inspiration in microeconomics and new institutional economics (Hirschleifer, 1977; 1987, ch. 9; Axelrod, 1984); still others refer by their 'evolutionary approach' to the application of ideas from the thermodynamic analysis of self-organising matter (Allen, 1981; 1988).
22 Nelson and Winter's (1982, 432) idea of calling their research programme 'evolutionary economics' is rather obvious. But in a certain sense the 'custom law' of the scientific community may have suggested to them that the intellectual property rights to this notion belonged to American institutionalism. And this tradition is based on methodological ideas quite different from the ones put forward by Nelson and Winter. The first to have used the notion of 'evolutionary economics' appears to have been Veblen (1898/1961, 76, 77, 79) and for some reason it became a

synonym for 'institutional economics', even if work in this field very seldom contains an explicit analysis of the evolution of institutions (see, e.g., the collection of papers on *Evolutionary Economics* by Tool, 1988). The organisational structures of American institutionalism help to explain the usage of words: The tradition from Veblen is maintained by the Association of Evolutionary Economics and its *Journal of Economic Issues*. The name of the association reflects Veblen's sketchy research programme but its practice is, of course, influenced by actual research practice which for long periods concentrated on a comparative-static analysis of institutions and on detailed empirical testing of hypotheses about individual institutions. For this reason one may say that institutionalism has lost its sole right to the notion of evolutionary economics if such a right had even existed. The emergence of a generalised notion of an evolutionary-economic process changed the situation and made it natural to define evolutionary economics in new ways. What is left is to define the different types of evolutionary economics and to analyse their interconnections and splits. This task has only been dealt with in a very rough way in the present book. Especially, I have not been able to incorporate the work of economists and sociologists based in Veblen's tradition of institutionalist. In this way I develop the practice of Nelson and Winter, and other 'neo-Schumpeterian' economists, which has been criticised by Foster (1991). I can, however, refer to Hodgson (1993) who has done much to reintroduce American institutionalism into the debates of modern evolutionary economics.

2. Schemes of punctuated evolution and jerky innovation

2.1. From railroadization to analytical schemes

In order to give a fair judgement of the applicability and adaptability of Schumpeter's conceptual structure, we can do no better than to consider the age of railway construction in the nineteenth century—both in its pioneering and its more mature stages.[1] The pioneering period was the time when the horse-driven mail coaches were outcompeted, railway towns mushroomed, financial schemes blossomed and failed, industries supplying and using the railways were set up, etc. The period of maturation was characterised by the routinisation of what earlier had been novelties and by the emergence of early forms of the modern corporation with its processes of rationalisation and control. More abstractly, it was an age of an irrevocable change of the routines of economic life which in many ways suggests an evolutionary-economic analysis.

Schumpeter has declared this process of 'railroadization'[2] to be the standard case of his evolutionary scheme,[3] a paradigmatic link between his analytic scheme and different real-life processes. This is underlined when Schumpeter (1939, 304) points out that this case demonstrates 'how railroad construction produces both prosperities and recessions ... and, in particular, simultaneous cycles of different span.' He urges that

> ... the reader should not fail to work this out again step by step. For railroadization is our standard example by which to illustrate the working of our model. ... [Many factors] combine to make the essential features of our evolutionary process more obvious in this than they are in any other case. More easily than in any other can the usual objections to our analysis be silenced by a simple reference to obvious facts. (Schumpeter, 1939, 304)

Such a conspicuous case[4] may help to focus many discussions in evolutionary economics. Especially it may help to develop Schumpeter's core scheme of evolutionary analysis. This may be called a punctualist scheme (Awan, 1986/91) since it emphasises that evolutionary processes are split up into epochs divided by short punctuations. In modern evolutionary biology the idea of punctuated equilibria is a minority view (Gould and Eldredge, 1977; Eldredge, 1989) which emphasises that long periods of stasis is punctuated by short periods of rapid evolutionary change. In Schumpeter's case the perspective appears to be different. His emphasis is put on the relatively long period during which a set of pioneering innovations leads to a basic transformation of the system of economic routines. Such periods of evolutionary change are punctuated by short periods of a relatively equilibrated system of routines. Therefore, I shall talk punctuated evolution rather than of punctuated equilibria.

In order to understand Schumpeter's evolutionary scheme, it is important to find out whether his punctualism concerns the character of the evolutionary process or the best way to study this process. In this context it is important to note that a scheme is basically a methodological device which may or may not give rise to theoretical statements. In other words, the scheme represents first of all a methodological punctualism and it will not necessarily lead to an ontological punctualism. The scheme suggests that in order to come to grips with the evolutionary process, we should start by considering it as divided into different segments or 'historical "individuals"' (Schumpeter, 1912/34, 58). In the end we may come to recognise that the evolutionary process may better be described in gradualist terms, just as Cuvier's account of biological evolution in terms of a series of catastrophic events was superseded by Darwinian gradualism. However, this is clearly not what Schumpeter thinks. He himself is suggesting a punctualist scheme as well as a punctualist theory of evolution. But the latter is historically specified: it is especially an unregulated capitalism which shows punctuated evolution.

Let us start by an attempt to sum up a Schumpeterian approach to central aspects of the process of railroadization (Schumpeter's treatment of this case[5] is presented in Andersen, 1993a). We would like to include several aspects of Schumpeter's treatment like: (1) the repeated contrasting between mail coaches and railways, (2) the emphasis on the large expenditure and the long periods of gestation of railway projects, (3) the subsequent transformation of all the data or routines of the economic system, (4) the disharmonious way in which changes take place, and (5) the final phase where routine business is more or less reestablished. These and other aspects of Schumpeter's treatment of railroadization are not only reflecting the historical facts but just as much his analytical scheme, his eyeglasses. This can be seen by a short orientation in the extremely rich literature on the great period of railroadization of the last century (a couple of UK examples are Simmons, 1978; Bagwell, 1974). Here a lot of other aspects of the case are developed. It is in a comparison with such contributions that the uniqueness and importance of Schumpeter's scheme begins to become clear.

Unfortunately, we also find out that it is not at all easy to express Schumpeter's account for the history of railroadization in precise terms. Especially, we are lacking a concept for the full application of an innovation in a given economic system. It is not at all easy to define such a concept in relation to an evolutionary process during which the innovation (like railway-based transport) as well as the overall system of economic routines is changing. Let us take the risk and apply a term from evolutionary and ecological biology, namely that of the 'carrying capacity' of the system with respect to a particular innovation (see chapter 3). In economic systems we are, however, basically dealing with the decision-makers' perceptions of the carrying capacity of the system with respect to certain profit and pricing assumptions, etc. If we at present assume an intuitive meaning to the concept, then we may make a picture of the process of railroadization which is simplified to the degree that it becomes a caricature:

1. We start with an old economic system where the capacity of mail-coach-based transport is close to the carrying capacity. In this situation finely tuned strategies of (simple or innovative) adaptation are dominant.
2. Then the railway is introduced as an innovation of mode of transportation. It takes a long time before the carrying capacity of the renewed system is reached because the ultimate carrying capacity is quite large and because the construction of this capacity requires a large amount of economic resources. In the meantime railway transportation influences some carrying capacities for other technologies and industries negatively (mail-coach transport) or positively (steel, mobile steam engines). Entrepreneurial strategies of implementing innovations in the different industries dominate in the beginning. *entrepreneurs in the beginning*
3. After several waves of construction and prosperity as well as recessions, all industries approach the new carrying capacities, including wide-spread railway transportation and zero long-distance transport by mail coaches. The successful firms are now following finely tuned strategies of risk-aversion, including some incremental innovation.
4. Then the automobile enters the economic scene, and the story starts once more.

The same story can be rephrased in a simplified form:

1. We start with equilibrated system of economic applications of routines, including the routines underlying mail-coach-based transport services. *innovation introduction*
2. Then the system is disturbed by the introduction of an innovation (railway-based transport services) with a large potential and with large resource needs.
3. A major reorganisation of the system of routine applications takes place through several cycles of economic expansion and recession. The end result is a (relatively) equilibrated state, partly based on the routines of railway transportation.
4. Then this system is disturbed by the introduction of a new innovation (automobiles) ...

When random
When in response to a performance gap

Schemes of punctuated evolution and jerky innovation 29

These formulations do not cover many important issues of the history of railroadization. Instead they reflect a highly stylised version of the case which may help to systematise much of the material. Furthermore, they lead us towards a more general formulation of Schumpeter's scheme of evolutionary analysis which could have been found by stripping his visionary accounts of the entrepreneur—manager interplay to their very backbones. The more radically we are able to proceed in this process of abstraction, the easier it will become to understand and develop Schumpeter's analytical scheme. Let us have a first try:

1. We start with a non-innovative 'circular flow' of economic life which follows given 'routines' that are upheld by strong preservative forces.
2. Then an irrevocable 'disturbance' of the 'circular flow' is created by the introduction of an 'innovation' by 'entrepreneurs'.
3. Finally a new, non-innovative state is obtained which includes a routinised version of what was originally an innovation; this is obtained by strong stabilising and selective forces.
4. The evolutionary process consists in a sequence of such steps.

Even this account might give unnecessary associations and constraints and thus restrict the analytical process. It might be helpful to let the process of abstraction go on and develop a general scheme with no explicit mentioning of economic conditions and economic agents. Such a scheme will primarily be applied on the study of the evolution of economic systems but it may also be used on other types of social evolution and even (for comparison) on biological systems. We arrive at:

(1) A system of routine behaviour
(2) is radically challenged by the innovative behaviour of the few,
(3) but sooner or later the equilibrating forces will establish a *new* system of routines.
(4) And then the story starts once more...

Through this procedure of abstraction, we are approaching Schumpeter's core scheme. However, it is necessary to develop the process of abstraction since even the last formulation has too many associations hidden behind the everyday words ('routine', 'radically challenged', 'innovative', 'few', 'equilibrating forces', 'new'). More seriously, Schumpeter's argument is still combining a descriptive scheme and theoretical propositions. On the descriptive side, we suppose that we can discern between two situations for the system of economically determined applications of routines. Either the application of the different routines shows no significant change during a certain period or we find that some routines (underlying, e.g., transportation) are applied relatively less while others are applied relatively more. In modern terms we may say that in the latter case there is a difference in the observed 'fitness' of the different routines (Winter, 1987). On the theoretical side, we make an interpretation of the stability and change in the 'fitness' of different routines. Here, Schumpeter's claim is that there are two opposing sets of forces which can explain the observed phenomena. On the one hand, we have a set of equilibrating forces which move the relative application of the different routines towards a state where they do not change significantly. On the other hand, we have the forces of (Schumpeterian) innovation which create new routines and thereby change the fitness of well-established routines.

The different formulations of Schumpeter's argument and scheme of analysis indicate an intricate mixture of descriptive scheme and theoretical claims, phenomena and underlying forces. It is not least this mixture which makes Schumpeter's (1991a, 230) ideas 'so refractory to mathematical formulations.' Even in the condensed form of the above formulations, we still see a mixture of methodological heuristics and theoretical claims about the evolutionary process. From a methodological point of view, it may be claimed that it is hopeless to study the real effects of innovation (railroadization) without a preliminary analysis which starts in the well-defined case where strong equilibrating forces are assumed to have created some sort of equilibrium in the pre-innovation (pre-railway) system of routine applications. From a theoretical point of view, one may, e.g., argue that the entrepreneurs' calculations about the profitability of radically new projects presuppose some stable reference price system which is most obviously ensured when the economic applications of other routines are near some sort of equilibrium. Thus, equilibrium may appear to be a preferable starting point both as a 'theoretical norm' (Schumpeter, 1939, 35-44) of the theorist and as a 'business normal' (pp. 3-6) that the innovative entrepreneurs can operate against. Both kinds of 'normals' help to make it easier to follow some of the many repercussions of the innovation. They also show why it is preferable to let the next major set of innovations (automobilization) start in an equilibrated system which includes the old (railway) routines. Methodologically it makes it easier to follow the evolution of innovation (automobile) applications and their effects. But what about reality? Is there, e.g., any necessary connection between the end of the age of railroadization and the coming of the age of automobilization? Schumpeter's answer appears to be: No, there is no necessary connection. But there is a lot of *ex post* evidence of interrelations. For example, railroadization did not remove the demand for horse traction as was initially believed. On the contrary, the railways increased the demand for horses since they were used, e.g., for short-distance transport to the railways (Thompson, 1976; cit. after Berg, 1980, 29 f.). This was the background for the norms of transportation in cities as well as in the country which the automobilization could relate to. Thus, the relative maturation of railroadization may have helped the subsequent wave of automobilization. But there are no grounds to claim a necessary relationship: we are basically dealing with an analytical scheme rather than a full-blown theory of economic evolution! This methodological character of the scheme becomes evident if we try to apply it to less conspicuous cases of innovation like 'a particular kind of sausage or toothbrush.' (Schumpeter, 1942/87, 132)

2.2. Measuring the application of routines

According to Schumpeter (1939, 87) the phenomenon of economic change has three different types of explanation: 'Economic change is due to External Factors, Growth, Innovation.' Wars and natural catastrophes are conspicuous examples of external factors which may be related to the waves in economic activity (p. 7 f.; see Goldstein, 1988). Growth due to population changes, saving and accumulation is a more theoretical factor which Schumpeter relates to certain types of expansion (1939, 82-84). However, innovation is clearly his central factor:

[N]othing can be more plain or even more trite common sense than the proposition that innovation, as conceived by us, is at the center of practically all the phenomena, difficulties, and problems of economic life in capitalist society and that they, as well as the extreme sensitiveness of capitalism to disturbance, would be absent if productive resources flowed—either in unvarying or continuously increasing quantities—every year through substantially the same channels toward substantially the same goals, or were prevented from doing so only by external influences. And however difficult it may turn out to be to develop that simple idea so far as to fit it for the task of coping with all the complex patterns with which it will have to be confronted, and however completely it may lose its simplicity on the way before us, it should never be forgotten that at the outset all we need to say to anyone who doubts is: Look around you! (Schumpeter, 1939, 87)

The study of the case of railroadization serves as such a look around. It gives some background for understanding Schumpeter's ideas of innovation and its role in the economic process. It relates to some of 'the phenomena, difficulties, and problems of economic life in capitalist society' but one can hardly say that it is able to demonstrate that innovation is the central cause behind 'practically all' relevant problems. It is not at all clear why difficulties and sensitivity are only related to innovation and not to the flow of economic resources 'through substantially the same channels toward substantially the same goals' but disturbed 'by external influences'. To study this proposition it is not enough to look around in an unprepared way; we must put on Schumpeterian eyeglasses and see whether the picture we arrive at is more convincing than alternative pictures (of, e.g., gradualistic evolution or exogenous shocks).

The formulations about the railway case immediately raise the question of what it is that evolves. This question appears to be answered by the above arguments: What evolves is the system of routine-based behaviour. However, in the present context we need some elaboration of this statement: What evolves is, in my opinion, the set of types of routines and the frequency of the economic application of the different types of routines. Thus we see that evolution can be described at two levels. On the one hand we have changes of the set of types of routines, the G-set. On the other hand we have changes of the observed (defined or tautological) 'fitness' of these routine types (see, e.g., Winter, 1987).[6] Those routine types that increase in frequency of application *relative* to others are said to be fitter or rather to have a higher observed fitness. If we have n types of routines, the development of their different fitness implies movements in the n-dimensional space of relative frequencies of routine applications, which we may call the F-space. If there is no supernormal and subnormal fitness, then there is no movement in the F-space. Within this set-up we are able to make a definition: Evolution in period t means either changes of the G-set (and thus changes of the F-space) or movements within the F-space. If none of these types of change have occurred, there has been no evolution in period t. This definition means that changes in *absolute* frequencies created by multiplying all the original frequencies with the same number are not evolution. This case of non-evolution appears to cover what Schumpeter calls Growth (and even a certain type of cycle).

The transformation of the discussion from the G-set to the F-space relates directly or indirectly to the difficult and controversial concept of 'fitness' which was introduced in relation to the above discussion of the F-space. Both biological species and economic routine types that increase in frequency of occurrence *relative* to others are said to be fitter or to have a higher observed fitness. This concept may have associations with Spencer's 'survival of the fittest', Darwin's 'natural selection' and the different schools of Social-Darwinism. But a concept is not explained by its uses and abuses. The concept of (relative) fitness may be applied for methodological reasons without the faintest normative overtones. In the present context the use of the term fitness is intended to suggest that the observed changes in the relative frequencies of routine-applications are not accidental. The idea of fitness gives a *possible* interpretation of the movement in the F-space of relative frequencies of routine-applications. If the relative frequency of the application of a certain routine increases, the routine is said to have a supernormal observed fitness; if it decreases, the observed fitness is subnormal. We may in a certain country in a certain period observe that railways expand and take over an increasing share of the transport services while mail coaches are squeezed out. By making such a description we have explained nothing. But the notion of observed fitness reminds us that it might be possible to develop some non-tautological propositions about the theoretically expected fitness in the concrete case which may in turn be compared with the empirically observed change in relative frequency. In this way the situations for the evolutionary biologist and economist are similar.

However, in economic life the routines are not built into individuals in the same way as in biological life. All economic decision-making involves implicit or explicit expectations about future situations, including speculations about the *ex ante* fitness of a given routine type and the possibility of imitating a routine type used by other individuals. Therefore, one may ask whether it is the railway routines or the firms of the railway business which show larger fitness. If we try to answer this question realistically, the analysis will immediately become extremely complex. Instead I will at present give a simplifying answer which will be developed into a somewhat more satisfactory answer in the subsequent chapters. The answer is that in the railway case and many other cases we see a tight binding between firms and routine types. Thus it is the application of the railway routines which gives the railway firms a larger fitness than mail coach firms. The intermediate situation (a firm covering both areas) is ruled out by Schumpeter since 'in general it is not the owner of stage-coaches who builds railways' (Schumpeter, 1912/34, 66). In modern terms we may say that there is, e.g., a kind of lock-in of the specialisation of mail-coach firms which determines their fate. Only the innovative entrepreneur has something like a real choice but as soon as his firm becomes a success, he tends to become locked-in like all the rest.

Schumpeter was not only concerned with partial questions but especially with the analysis of the system of routine-applications as a whole and, in my interpretation, with general movements in the F-space. It is through such movements that an innovation may develop to system-wide importance. In relation to this overall question we would like to see the applications of a given routine type relative to all other routine applications. But we may start by calculating the absolute fitness of the individual routine types and afterwards try to exclude general Growth as well as a (very special) form of cycle from the area of our attention. To see how this is done we may consider the 'absolute fitness' of the jth class of routines in the economy, $f_j^a(t)$:

$$f_j^a(t) = \frac{N_j(t+1)}{N_j(t)}, \tag{2.1}$$

where t and $t+1$ are two subsequent points of time and $N_j(t)$ is the frequency of the application of routine class j as measured by its embodiment in stocks of commodities or productive capacities at time t. This index of absolute fitness is simply 1 plus the relative growth rate. If the absolute fitness of j is 1, then there is no change in the frequency of application between the two periods. If it is above 1, then the number of applications is increasing. If it is below 1, then the number is decreasing. If we compare the vector of all absolute fitness ($j = 1, ..., n$) at two points of time, we say that the System has shown Growth but not Evolution if $f^a(t+1) = kf^a(t)$, where k is an arbitrary constant (positive or negative). In the very improbable case that $k(t)$ is an oscillating function of time, we will say that the system shows non-evolutionary cycles.

At the system level the notion of fitness as change in relative frequencies removes from the very start the non-evolutionary cases of growth and simple cycles. But this service is hardly sufficient to justify the use of relative rather than absolute fitness. However, the notion of relative fitness increases in importance if we also apply it at a lower-than-system level. Here it makes it easier to discuss the interaction between different but related routine types like rabbit types and types of transport routines. In these cases we are really subdividing the vector of absolute fitness, $f^a(t)$, into aggregates of connected routine-applications which are then compared with the change in the absolute frequencies of the component parts (all transport services and railway-based transport services). In this case we make a distinction between larger classes or groups of routines, j, each of which consists of one or more routine types, i. Thus, we presuppose a double classification of all the routines of the economy. To illustrate the problems related to this task we may delimit the area of interest to the types of routine-applications which are reflected in the commodities of the economic system. The data can be collected by making an analytically relevant inventory of all the commodities stored somewhere in the economy at a particular point of time, t (see a related example in Winter, 1987, 614 f.). In principle each individual instance (implementation) of a commodity type is unique. But in evolutionary analysis we would like to make (at least) a double classification of the individual commodity instance. To begin with it should be recorded as belonging to a particular commodity type, i, and to a larger commodity group, j. This is done for all the 'commodities' in the economy (including physical capital). At a later point of time, $t + 1$, we make a similar inventory. On this basis we calculate the absolute fitness of commodity group j. Finally, we calculate, $f_{ji}(t)$, the relative fitness of commodity type i which belongs to commodity group j:

$$f_{ji}(t) = \frac{N_i(t+1)}{N_i(t)} \bigg/ \frac{N_j(t+1)}{Nj(t)} \tag{2.2}$$

To the extent that j is defining a group of is which are subject to more or less homogeneous selection pressures, equation 2.2 becomes especially fruitful for the purposes of partial analysis.[7]

Absolute fitness, $f_j^a(t)$, and relative fitness, $f_{ji}(t)$, are notions which will play an important role both as descriptions of facts, and in defining areas for theoretical explanation. For this reason it is important to note some serious problems about their use which are again and again pushed to the forefront in Schumpeter's works and in Schumpeterian analyses.

First of all, the definition cannot be used to assign a fitness to a Schumpeterian innovation, i.e., a new commodity type where, e.g., $N_i(t) = 0$ and $N_i(t+1) > 0$ (we must not divide by zero). Schumpeter closes any escape to the infinitesimal method by emphasising that in the case of innovation we are facing an inevitable jump. However, we may have some theoretical ideas about the potential carrying capacity of the system with respect to the new commodity type as well the rapidity with which this level may be approached. Furthermore, the idea of a potential 'fitness' of an original innovation may also be related to the *ex ante* judgements of the entrepreneur (and his counterparts in banks, etc.). The entrepreneur is, so to say, making a theory of the fitness of the individual innovation. In any case, the notion has nothing to do with observed fitness.

Second, we have the case where the economic application of a routine dies out, i.e., $N_i(t+1) = 0$ and the fitness is zero. Here we cannot be absolutely sure that we are facing an irreversible situation (mail-coach transport may become popular for purposes of amusement). However, in economic life there is no book of blueprints which makes commodities with zero frequency a living reality in decision-making. For all practical purposes we may consider a disappearance as irrevocable and the emergence of a somehow similar commodity to be an innovation.

Third, the definitions are dependent on the availability of an adequate taxonomic system which is seldom available. While the biologists have discussed taxonomic problems for hundreds of years and gradually are changing their taxonomies to reflect their evolutionary results and theories, nothing comparable has happened inside economics. To the extent that the economic actors and the economic system of naming of commodities are subject to evolutionary pressures, we may be provided with relevant information. But since exact information may help competitors we should not be too hopeful. Furthermore, the taxonomic systems of the statistical services of the different nations are very conservative. In other words, there is no guarantee that we can take evolutionary adequate *i*s and *j*s out of the given statistics.

Fourth, even if many kinds of statistical heterogeneity are unwelcome, it should be emphasised that the kind of evolutionary process we are studying implies much variance at almost any level of aggregation. Thus we should not expect the *i*s and *j*s of an ideal taxonomy to be homogeneous. In the case of transportation *j* may be the total capacity of transportation, *i* may be mail coach capacity and *i* + 1 may be railway capacity. We may also work at a lower level of aggregation where *j* is miles of railway track, *i* is miles made by wrought-iron rails and *i* + 1 is miles made by steel rails. But steel-based tracks may also be considered as a heterogeneous commodity group. We may proceed in this way until we come to changes which are not easily caught by statistical analysis. Here we may turn to one of Schumpeter's many comments on the dangers of statistical analysis which relates to agricultural commodities:

We gather from historical indications that even the oldest and most ordinary articles of consumption, such as meat and wine, are quite different now from what they were even 100 years ago. But in most cases we have no means of measuring the change. (Schumpeter, 1939, 484)

These remarks give a kind of conclusion on the discussion of the difficulties involved in the special type of heterogeneity and irrevocable change which characterise evolutionary processes. In some situations we know that some kind of evolution has taken place but must (for the moment) leave the question. At other times we may proceed and study the subject matter thoroughly. Most often we may suppose that the quality changes in meat and wine are so slow and have so little to do with the overall processes under study that it can be assumed that no change in meat- and wine-related routines have taken place.

The list of difficulties related to the notion of fitness is far from complete. The most important difficulty is, perhaps, that the notion lends itself easily to tautological modes of argument: 'routine-x is fitting into a given selection environment because it shows a high fitness'. The difficulties involved in the idea of fitness may explain the tautological character of some evolutionary arguments which has been severely criticised by Popper (1972). But many discussants argue convincingly that it is possible to avoid tautology if the evolutionary arguments are aware of the crucial distinction between theoretical and observed (tautological) fitness. In the present interpretation the naming of observed fitness implies that the observations are heuristically expected to relate to the evaluation of the characteristics of the routines by a selection environment. Such a heuristic may be applied by the theorist as well as the economic decision-maker. Therefore, we appear to have three notions of fitness: theoretical fitness, *ex ante* fitness representing economic decision-makers' expectations, and observed or *ex post* fitness (see section 3.4.2).

2.3. The scheme of punctuated evolution

We are now ready for yet another step in the process of abstraction which started from the case of railroadization and ended in a verbal version of Schumpeter's scheme of punctuated evolution (section 2.1). The basic idea is to give a formalised description of the scheme. Instead of having a scheme of managers and entrepreneurs, we define their function in the process of evolutionary change in descriptive terms. Instead of innovative and equilibrating forces, I will try to describe two abstract operators with specific functions in the evolutionary process: an α-operator (propelling evolution) and a β-operator (working towards a non-evolutionary state). They operate on the system of routine-based behaviour and their action can be described at two levels. On the one hand, we have changes of the set of types[8] of routines, the G-set. On the other hand, we have changes of the frequencies of these routine types which are described in terms of movements in the F-space. If we have n types of routines, the development of their frequencies implies movements in the n-dimensional space. If there is no supernormal and subnormal fitness, then there is no movement in the F-space. Within this set-up we are able to make a definition: Evolution in period t means either changes of the G-set (and thus changes of the F-space) or movements within the F-space. If none of these types of changes have occurred, there has been no evolution in period t.[9] Since balanced growth gives no change in (relative) frequencies, it shows no evolution. This case of non-evolution seems to cover what Schumpeter calls Growth (and even a certain type of cycle[10]).

Table 2.1. The function of the operators of evolution.

	G-set	F-space
α-operator	Introduction of new routine types	Creation of disequilibrium
β-operator	Possible removal of old routine types	Movement towards equilibrium

On the basis of the preliminary definition of evolution we are able to make our definition of the two 'forces' in relation to the evolutionary process more precise: the creator of evolutionary change, the α-operator, and the creator of movement towards equilibrium frequencies of the different routines, the β-operator (see table 2.1). The α-operator is intuitively related to the mechanisms of variety-creation in the routine system while the β-operator is related to the selective mechanisms involved in the ideas of (non-creative) competition and cooperative economic interaction.[11]

The forces may be considered to function in parallel just like in biological evolution. If one of the forces comes to dominate totally, we will have no evolution: either we will end in stasis because of the dominant β-operator or we will have 'chaos' because of the dominant α-operator. But between these extremes there are many evolutionary possibilities. Some of them will take the form of waves of evolution where the two operators are in turn dominant. One possibility is that the evolution-creating operator is at first auto-catalytic so that one change in the G-set gives rise to another; but as radical disequilibria are spreading and as more and more of the flexible resources are bound, the conditions of the α-operator become more and more difficult and the β-operator becomes dominant. This operator selects the routines in a way which implies that their relative frequencies of application move towards their carrying capacities where further change has stopped and where some of the original routines may have a zero frequency of application (routines which are extinct with respect to the economic system). But before the system comes to this state, the inhibition of the α-operator may have weakened and thus the system may never come to its equilibrium state.

This picture of two evolutionary forces (of certain types of variety-creation and selection) working more or less in parallel but with some interconnections is probably well-suited for economic evolution while biologists emphasise the (near) total independence of the mechanisms of variety-creation and selection. But the picture is not at all easy to handle analytically. The reason is, of course, that we have not defined what it means to be somewhat closer and somewhat further away from equilibrium in the F-space. Probably this cannot be done in a fully satisfactory way as long as we stick to very general definitions in the present chapter. This may be one of the reasons why Schumpeter preferred to deal sequentially with the forces: If the β-operator has created equilibrium, then it is much more easy to handle the α-operator. If the α-operator stops its work after a while, then it is much more easy to study the functioning of the β-operator. We will now turn to this sequential procedure which is clearly reflected in the above Schumpeter-like formulations of the stylised case of railroadization as well as of economic evolution in general. But we should keep in mind the possibility of the parallel working of the α- and β-operator.

Let us call the state of the economic system where no evolution takes place the Σ-state (or the non-evolving 'circular flow') and the state where evolution takes place the Δ-state ('economic development' or 'economic evolution'). In this setting we may talk in terms of restricted versions of the operators which shift the state of the economic system between its two basic states. The α-operator changes the state of the system from a Σ-state to a Δ-state; the β-operator changes the state of the system from a Δ-state to a Σ-state.[12] In this way we may try to describe Schumpeter's most abstract scheme in the following terms:

Schumpeter's scheme of evolutionary analysis may be described in terms of:
(1) a set of types routines, the G-set, and a related space of frequencies of the economic application of the routine types relative to each other, the F-space;
(2) a Σ-state which is defined by the lack of change of the G-set and the lack of movement within the F-space;
(3) a Δ-state which is defined by change of the G-set or movement within the F-space;
(4) an α-operator which takes a Σ-state and returns a Δ-state,
(5) a β-operator takes a Δ-state and returns a Σ-state.

Compared with the schemes above, the new formulation is difficult to read. But the present formulation of Schumpeter's scheme helps to emphasise that the definition of G-set and the F-space plays a central role in the whole argument. If it is not possible to identify a sort of routines which are classifiable and replicable in an economically relevant way, the whole argument breaks down. But if we have such routines, the Σ-state is defined by the lack of evolution, i.e., by a peculiar descriptive type of 'general equilibrium of routine frequencies'. This concept of equilibrium is clearly different from the concept of the general equilibrium of the normal economic variables (prices and quantities of commodities) since there may exist changes in the economic variables (general inflation, growth and perhaps other cases) in periods where the system stays in its Σ-state.

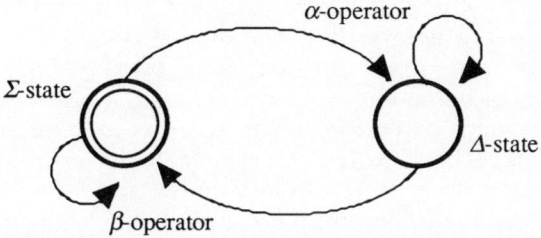

Figure 2.1. Schumpeter's scheme of evolutionary analysis.

The sequential character of Schumpeter's scheme is summarised in figure 2.1 Here a certain asymmetry between the two states of the system is indicated. The Σ-state is marked with a double circle indicating that the evolutionary process may, in principle, settle in this state which thus is a (semi-) terminal state of the evolution of the system. On the other hand, the Δ-state is defined as a non-terminal state of the evolutionary process. This means that the α-operator may create a Δ-state while the α-operator must return the system to its Σ-state. As discussed above, this is a consequence of Schumpeter's methodological considerations and should not necessarily be seen as a theory of the evolutionary process (a problem which will be taken up later). Another problem with figure 2.1 is that the system never returns to the same Σ-state after it has been left as a result of the α-operator which is defined in terms of changes of the G-set and F-space. In other words, this operator introduces an element of irrevocability into the process (which is the reason for calling it evolutionary). For this reason the evolutionary process is, perhaps, better illustrated by figure 2.2.

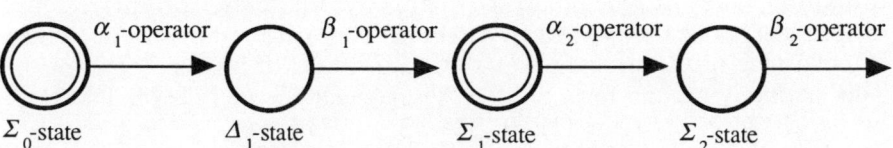

Figure 2.2. Schumpeterian evolution as an irrevocable process (time from left to right).

The way 'history' intrudes into the argument is quite difficult. A first idea of what is going on may be developed by thinking of a succession of 'technological' or 'techno-economic' revolutions. The Σ_0-state is some initial state with a (more or less) well-defined state-of-the-art widely applied throughout the economic system. The α_1-operator is the introduction of railways into the economic system. This creates a turbulent period but sooner or later railway-based are part of business-as-usual in the Σ_1-state. There is a marked difference between the routines of G_0 and G_1 but we have a basis for the operation of the α_2-operator. In which way we may say that α_1-operator $\neq \alpha_2$-operator cannot be dealt with here. But we have already had a glimpse of Schumpeter's relationship to the Historical School of economics.

Another important extension of the scheme concerns the distinction between what have been called 'incremental innovations' and 'radical innovations' or, better, 'microinnovations' and 'macroinnovations'. (Mokyr, 1990) We may interpret the α-operator as being connected only to 'macroinnovations'. This leaves most innovations to the realm of the β-operator which is, however, covering many non-innovatory processes. For this reason we would like to distinguish between the incrementally innovatory β_α-operator and the non-innovatory β_β-operator. To develop this idea, we need to define the set of routines which is changed by the β_α-operator, and the G_β-set which records the introduction of microinnovations while the G_α-set deals with macroinnovations. The problem is to decide whether an innovation belongs to the one or the other of these sets. But in principle we have a way of distinguishing between some of Schumpeter's core arguments which deal with the α-operator and much of the new evolutionary economics which deals with the β_α-operator.

2.4. Punctuated evolution and Schumpeter's 1910-theses

It is now time to turn to Schumpeter's original formulations and confront them with my reconstruction, as well as with a few contrasting formulations. In this way we have a check whether the present reconstruction is drifting too far away from Schumpeter's mode of thinking. One of the places where Schumpeter formulates his evolutionary scheme most clearly is in his very first and totally neglected summarising of the structure and the main results of his type of analysis. This summary has the form of nine theses which are included in his paper on the *Essence of Crises* (1910, 324-325, in German).[13] The theses are clearly made as part of his preparation for the writing up of *The Theory of Economic Development* (1912). They are expressed in terms like 'statics' and 'dynamics' which were later changed into 'circular flow' and 'economic development' or 'economic evolution'.[14] There is no concept for the overall evolutionary process in terms of a series of punctuated evolutionary periods.

With these remarks in mind we can turn to Schumpeter who recommends

... to express the basic ideas of our expositions in the following theses:

First, the economic processes divide into two different and also in practice clearly discernible classes: static and dynamic.

Second, the latter constitutes the pure economic evolution , i.e. those changes in the model of the economy that arise from itself.

Third, the economic evolution is essentially a disturbance of the static equilibrium of the economy.

Fourth, this disturbance provokes a reaction in the static masses of the economy, namely a movement towards a new state of equilibrium.

Fifth, this process of statisation necessarily makes an end to every concrete phase of evolution and leads to a reorganisation of the value- and price system of the economy and a general 'liquidation'.

Sixth, these propositions explain the phenomenon which is popularly called the shifting of prosperity and depression.

Seventh, during the process of statisation and especially in the moment of its set in, those breakdowns can arise that we $\kappa\alpha\tau\ \varepsilon\xi o\chi\eta v$ [by way of eminence] call economic crises and which make the process an 'abnormal' one.

Eighth, the economy, and certainly the static one, is, furthermore, exposed to contingent disturbances which can when sufficiently important cause such crises.

Ninth, however, these [crises] constitutes no further problem, on the contrary they are immediately understandable. They are in no important way a uniform phenomenon, do not carry any deeper common characteristics and do not arise from some necessity given by the economy or a particular organisational form of it. Against them is the ruling opinion that crises occur precisely when a larger disturbance sets in somewhere in the economy not only correct but also fully exhaustive. (Schumpeter, 1910, 324 f., my transl.)

The theses are rather compressed, but the general structure is relatively simple: The first five theses summarise Schumpeter's overall analytic scheme without mentioning explicitly innovation and entrepreneurship: they are working at a higher level of abstraction. The next two theses (the sixth and seventh) postulate that the scheme defined by the first five theses can be used for the explanation of business cycles and crises to the extent that they are generated endogenously within the economic system. The final two theses (the eight and ninth) point out that crises may also have exogenous causes but that this phenomenon is, in Schumpeter's eyes, of no particular interest to economic theory.

Our first question is whether the first five theses may be translated into the terms of the formalised scheme of punctuated evolution in a way which does not disturb the structure of the argument. Let us have a try:

1. The economic processes divide in theory and practice into two sharply distinguished classes: those related to the Σ-state and those related to the Δ-state.
2. The latter we call pure economic evolution since they concern changes in the model structure (ultimately defined by changes in the G-set) that arise as the consequence of the α-operator which is internal to the economic system.
3. Economic evolution created by the α-operator is defined as a disturbance of the equilibrium position in the F-space (and the corresponding Σ-state).

4. The creation of a Δ-state induces the actions of the β-operator which determines a movement towards a new equilibrium position in a new F-space.
5. This change process determined by the β-operator will bring the system from its Δ-state to its new Σ-state, and the latter presupposes a total reorganisation of the value- and price system which was related to the original Σ-state as well as a wide-spread 'liquidation' of economic activities bound to the routines which are made obsolete in the new Σ-state.

This translation from Schumpeter's theses to the terms of the formalised scheme has not been made in a purely mechanical way and it raises several questions (including the way in which the α-operator can be considered to be internal to the economic system). It is, however, more adequate than a reformulation in terms of a Walrasian scheme of comparative-static which may lead to the following formulation of the first five theses:

1. We may discern between general economic equilibrium and disequilibrium.
2. Disequilibrium is created by purely non-economic factors, by an exogenous change of the parameters of the economic system.
3. Disequilibrium is a disturbance of the given allocation of resources.
4. This disturbance creates a process of arbitrage towards a new equilibrium.
5. This process will overcome disequilibrium and create a new equilibrated system of prices and allocation of resources.

The difference between the Walrasian and the Schumpeterian formulations is most obvious with respect to the second of the 1910-theses. In the Walrasian formulations the disruptive innovation is considered as exogenous and treated in exactly the same way as any other kind of disturbance, from minor changes in tastes to major catastrophes. In all these cases we have an exogenous change which has nothing to do with the pure analysis of economic life which is only related to the consequent reestablishment of equilibrium. This is clearly not what Schumpeter means: both the α-operator and the β-operator are considered to be endogenous forces. This gives further differences with respect to the 'translations' of the fourth and the fifth thesis. The Walrasian formulations cannot cope with terms like 'provokes' and 'liquidation'. To Walras 'economic life is essentially passive and merely adapts itself to the natural and social influences' (Schumpeter, 1937/51, 159). To Schumpeter things are not that smooth. There are also counter-forces which make the process of adaptation to new routines a painful process. Furthermore, it is these forces which ensure that innovations are not from the start introduced as soon as a possibility appears. In the

Walrasian scheme such possibilities must be exploited. In the Schumpeterian scheme they may in some situations be exploited by innovative entrepreneurs.

The arguments against a simple Walrasian interpretation of the theses provide, of course, no evidence for my own interpretation. Nevertheless, they underline, e.g., the importance of the routine concept which may reflect the 'stickiness' of economic life and the difficulties of its process of transformation. This problem becomes even more important when we come to the last four of Schumpeter's theses. They help to underline the broader context within which Schumpeter's work must be seen. This problem becomes especially salient when we come to theses 6-7 which may roughly be interpreted in the following way:

6. Economic 'prosperities' and 'depressions' may be connected to, on the one hand, the creation of a Δ-state by the α-operator and, on the other hand, the recreation of a Σ-state by the β-operator.
7. The action of the β-operator and especially its set in (with full force) is related to economic breakdowns and abnormal processes ('crises').
8. 'Crises' may also be created by exogenous shocks
9. but this phenomenon is quite heterogeneous, of little interest to analysis and not very important.

The punctuated process of economic life may not stand out sharply from the above discussion of the Schumpeter—Walras relationship but it clearly does so from the theses. But at least in the evolutionary scheme of the young Schumpeter, a wave of evolution is followed by a period where evolutionary change has come to a halt. This characteristic may have been developed for methodological rather than ontological reasons but it is clear from some of the formulations that Schumpeter thinks that the broad introduction of new innovations is connected to periods of 'prosperity' while their fight with the old routines and their full implementation are connected to periods of 'depression'. Furthermore, we see (in other parts of Schumpeter, 1910) that between 'depression' and the next 'prosperity' there must be a period of equilibrium (with no movements in what is here called the F-space). This is an idea which makes it practically impossible to translate Schumpeter's theses into a Darwinian type of discourse since this is emphasising the gradual and continuous march of evolution. Nevertheless, let us try to formulate the biological scheme of punctuated equilibria in the same general form as the first five theses:

1. The routine system of economic life shows long periods characterised by equilibrium and short periods of disequilibrium.
2. It is in the latter periods that real change in the routines of economic life occurs and these periods are probably initiated by

exogenous forces since equilibrated economic systems of routines are quite stable.
3. A period of rapid change is essentially a punctuation of a long-term equilibrium period.
4. This punctuation provokes a process of selection between a wide range of newly emerged types of routines and this process moves relatively rapidly towards a new equilibrium situation.
5. This selection process creates a new price system and causes a great many types of routine to become extinct.

This transformation shows clearly that even in relation to the most near-standing evolutionary biologists, there are far too many disanalogies to make a direct translation. Even with some bending of Schumpeter's theses, many obvious formulations put forward by the punctualist interpretation have no counterpart in Schumpeter's theses. Furthermore, there is a formulation in Schumpeter ('changes in the model of the economy that arise from itself') which sounds mysterious in the punctualist interpretation. How can the forces operating within long periods of evolutionary stasis by themselves create sudden punctuations?

For the new evolutionary economics there are other but even larger problems with Schumpeter's 1910-theses. The main problem is probably the assumed equilibrated state between each cluster of (Schumpeterian) innovations. But also coupling of the process of evolutionary change to the macroeconomic business cycles suggests many counter hypotheses. A normal reaction to such problems is to turn away from the macroevolutionary phenomena and to concentrate on Schumpeter's contribution to microevolutionary analysis. There is, however, a close relationship between the two and no long-term 'dialogue' between new evolutionary economics and Schumpeter is conceivable without some sort translation of his overall scheme into a language understandable by modern researchers. Some attempts do so are presented in chapter 5.

2.5. Economic paradigms and evolution[15]

In the previous sections Schumpeterian entrepreneurs and managers became transformed into abstract operators (section 2.3). It is now time to rethink the 'thing' which keeps Schumpeterian managers and entrepreneurs apart, namely the difficulty of introducing an innovation ('new combination') into the realm of normal economic decision-making. The most difficult type of a new combination is probably one which must be characterised as combining all of Schumpeter's five types of novelty: novelty with respect to product, process, input, region and organisational form. Then the difficulty decreases until we reach the gradual type of change which is not included in the concept of Schumpeterian new combinations or innovations. At this level it becomes practically and theoretically impossible to discern between managers and entrepreneurs (and to use the other Schumpeterian dichotomies). But it is exactly this gradual or incremental type of innovation which constitutes the bulk of what is today being studied under the heading of 'technological innovation' and which appears analogous to the kind of change studied within standard neo-Darwinian biology. There is a natural tendency to try to approach all kinds of evolution from this gradualist perspective but at the same time Schumpeter appears more and more obscure and old-fashioned. Even the parts of Schumpeter's work which are still supposed to be of relevance are gradually lost.

In the following sections it is argued that modern analysis may help to understand the phenomenon of the two modes of behaviour in terms of the lock-in into sticky or even inflexible conduct which is once in a while broken by the implementation of more profitable alternatives. This idea has some similarities with the notions of scientific paradigms and research programmes within the theory of science (Kuhn, 1962/70; Lakatos, 1970), ideas which were already present in Schumpeter's general evolutionary conception. The solution is, however, not identical with several other contributions to the theory of technical change and economic evolution. First of all, Schumpeter is clearly interested in 'economic paradigms' related to economic decision-making rather than directly in, e.g., 'technological paradigms' (Dosi, 1982; 1984). Second, the whole idea of entrepreneurs and managers making 'combinations' suggests that the firm should be considered as a black-box which is defined through its interfaces with its customers and its input suppliers rather than as a complex organisation living its own life.[16] Similarly, economic paradigms should not only be considered as means of intra-organisational coordination (or coordination between 'societies' of producers of technological knowledge) but rather as means of coordination between groups of sellers and buyers, or producers and users, of specific types of commodities and productive services. Such paradigms are shared specifications of typical interfaces between two parties. If such interfaces are simplified and standardised, the information needs of the parties will be delimited. However, successful entrepreneurship presupposes an information-rich interaction and thereby presupposes a period of non-standardised interfaces. In the following sections these two conflicting principles are developed in order to clarify several controversial issues within Schumpeter's theory of innovation and evolution.

The study of evolutionary processes in relation to a market-based economic system must deal with two related issues: the semi-stability of some elements of the system and the irreversible changes which these elements nevertheless undergo. Technology and the way it is coupled with the economic system may give clues for defining semi-stable characteristics which mutate and evolve. The recent discussion on 'technological paradigms' (see Dosi, 1982; 1984; 1988) opens up this perspective. The present discussion represents an attempt to modify and extend the notion of technological paradigms in a way which may increase its usefulness as a base for evolutionary studies. The main change which is proposed is to include more explicitly the interactions between the entrepreneurs and managers (collectively called 'producers') as well as customers ('users') of commodities. The 'paradigm' is seen as a mutually agreed definition of the producer-user interface which partly takes the form of specifications of the commodities to be delivered. Such specifications tend to become generalised to groups of producers and users of a commodity and they are difficult to change. A Schumpeterian entrepreneur is dealing with the establishment of new interfaces. but even the old and well-established interfaces play a crucial role for his functioning. A manager takes all interfaces for granted in his decision-making, but even to him there may be possibilities of gradual technological improvements within the framework of the given interfaces. This is what makes him different from the entrepreneur who breaks with the basic specifications in order to form a new group of users and producers (starting from his own enterprise).

The proposed conception of the activities within economic paradigms is shifting the focus away from Dosi's focus on a 'community of engineers' involved in the perfection and administration of a particular technology. Instead it focuses on the 'community of producers and users' of a given commodity and their influence on the related technology. The engineers of the producers and users do not form a homogeneous 'community', and many economic considerations are clearly involved. This is one of the reasons why I prefer to talk of 'economic paradigms' instead of 'technological paradigms'. I hope the two notions will prove to be at least partly complementary. The notion of 'techno-economic paradigms' (Perez, 1983; Freeman and Perez, 1988) may be considered as relating to a very special and all-encompassing type of 'economic paradigms', macro-paradigms which relate to whole Kondratiev waves. The present microscopic notion of economic paradigms appears to be much simpler, but it may nevertheless be difficult to grasp. The problems lie partly in conceptual issues, partly in the fact that the dominant tradition in the study of innovation and evolution takes its point of departure in single firms or industries in a relatively unstructured environment, and not in the interrelations between specific firms.

In their behavioural approach to the firm, Simon and others have developed a framework which points towards phenomena much like technological and economic paradigms. From this viewpoint paradigms may be considered as a kind of fixed-points or 'signposts' which guide technological development of a more routine-like type, as necessary guides of firms living in a world of bounded rationality. Such an approach might cover Dosi's ideas but not the micro-oriented economic paradigms. What is needed is, in my opinion, a complementary approach which emphasises the inter-organisational aspects of paradigms. Here the need is not so much Williamson's (1985) transaction-cost approach as elements of the later Austrian school's more fundamental approach to the evolution of knowledge accumulation and use in a complex system of economic transactions (especially as developed by Hayek, 1948). From this latter approach, paradigms may be considered as means of delimiting the needs of information flows between the actors, thereby providing a major precondition for the existence of complex techno-economic systems.[17]

In my view the interplay between producers and users has a much more fundamental role to play in an encompassing definition of technological paradigms.[18] Therefore, I define an economic paradigm as a mutually accepted and stable specification of the interface between producers and users of a complex type of artefact. This artefact or commodity is mainly understood in terms of major functions *vis-à-vis* the users, but normally a few specifications of materials etc. are also included. All other information of, e.g., the production process of the artefact or its uses is irrelevant to the interface. But the interface has clear implications for the concept of 'progress' for the producers as well as the users. To the users the relevant problems, procedures and knowledge base concern the relation between the price and the performance parameters which may be part of the interface. To the producers the relevant problems, procedures and knowledge base concern the efficient production of the given artefact/commodity and the profits derived from improving it according to the performance parameters. Then, we will call a 'technological trajectory' the direction of advance (from the viewpoint of the producers) within the framework created by the given interface.

The definitions show how I have placed the source of behavioural stability in the interface between two sets of actors engaged in exchange, an approach which is totally absent in Dosi's definitions. The stable thing is the basic design of the commodity to which the positive and the negative heuristics adapt, and to change the interface is difficult because it is rooted in the norms and organisations of the two parties. This conception is not easy to discuss on the background of the approach of the behavioural school where you normally conceive the firm in a more or less given environment. From this viewpoint economic paradigms may be seen as an internal characteristic of each organisation which tries to orient and adapt itself in a highly complex and uncertain world. The present problem is another one: How and why does the firm and its 'local' environment become structured in a relatively simple way which eases some of the information problems? Or, more fundamentally: How and why is the techno-economic world knowable?

The first part of the answer is that this world is characterised by a relatively high level of reproducibility, of routines and stable interfaces which make it possible for the agents to accumulate experience which functions as usable knowledge. The second part of the answer is that the economic and technological agents are forced to and try (partly inadvertently) to organise the techno-economic world in this knowable way. The basic problem is one of coping with complexity: a user might question the construction of one, or a few, of the inputs into his production process and might be involved in the development of an input product with new characteristics. However, the principle of procedural rationality tells us that in a complex intra- and inter-organisational network everything cannot be changed at the same time without creating 'chaos'.

For this reason it is not easy to stick to the picture of knowledge accumulation within a firm in a (relatively) unstructured environment. It becomes more and more clear that the firm is, in itself, characterised by a relatively limited ability to uphold a stock of usable knowledge (and skills). The firm's knowledge is to a large extent knowledge of a limited part of its environment ('localised knowledge'). This knowledge is only powerful as part of a system of techno-economic knowledge based on an extended division of labour between different knowledge-holding units. This viewpoint has some similarity to the interpretation of the market mechanism by the Austrian school of economists. This interpretation was developed by Hayek's (1948, 51) pointing at the problem of the 'division of knowledge' as 'the really central problem of economics as a social science.' Hayek has stated this central question as follows: 'How can the combination of fragments of knowledge existing in different minds bring about results which, if they were to be brought about deliberately, would require a knowledge on the part of the directing mind which no single person can possess?' (p. 54)

This question does not arise in the Walrasian world of perfect knowledge and infinite computational abilities of the economic actors. Under such conditions the many economic actors and their market interaction can logically be replaced by a central authority. This is not the case in Simon's world, but he is so concerned with the decisions of the individual firms in a given environment that he only loosely formulates 'the problem of knowledge'. However, in Hayek's world the decision-making has to take into account that (1948) 'the knowledge of the circumstances of which we must make use never exists in concentrated or integrated form, but solely as the dispersed bits of incomplete and frequently contradictory knowledge which all the separate individuals possess. The economic problem is thus ... a problem of the utilisation of knowledge not given to anyone in its totality.'

In this perspective, Hayek points to a new interpretation of the institutional set-up of the economy. The most obvious example is, of course, the price system of a market economy:

The most significant fact about this system is the economy of knowledge with which it operates, or how little the individual participants need to know in order to be able to take the right action. In abbreviated form, by a kind of symbol, only the most essential information is passed on, and passed on only to those concerned. (Hayek, 1948, 86)

At its best the price system serves to protect the decision-makers from the distractions of the enormous total amount of knowledge. It serves to reduce their attention to a localised and computationally manageable amount of information. In order to serve this function the set of prices should not change too much. The reason for this conservatism is also that localised pricing is the outcome of a long process of trial and error which, according to the Austrian outlook, is not easily reconstructed. What a difference between this view and the view of Walras who appears to consider the reconstruction of the price system to be an afternoon's work through the process of *tatônnement* headed by an imaginary auctioneer. The difference becomes even greater when it is made clear that the Walrasian world is based on a fixed set of types of commodities while the Austrian conception allows for changes in the quality and types of commodities as well as of quantities and prices. Here the need for 'economy of knowledge' is by no means confined to economic calculations and decisions. The engineers also rely on the well-defined and simple characteristics of the components and materials supplied for their construction of artefacts.

2.6. Two principles of interface design

2.6.1. The principle of commodity abstraction

Relatively stable and simple interrelations is of paramount importance for the functioning of the economic system. At the same time, the acts of the Schumpeterian entrepreneurs are critically dependent on the willingness of others to offer or accept new products. These points can be developed into a principle of the design of interfaces between groups of firms, as well as the necessity that the evolutionary process breaks this principle. In this way we are able to bring managers and entrepreneurs back into the picture. The principle of commodity abstraction minimises the need for information flows and may help to define the background of the behaviour of Schumpeterian managers. The breaking of this principle is what characterises the Schumpeterian entrepreneurs.

To understand this situation we may look back into an earlier age of techno-economic development where many of the simplest commodities could not be trusted except when dealing with agents heavily dependent upon retaining the local goodwill associated with their firm, i.e., to a situation with some similarities to the one modelled through the Trader's Dilemma in chapter 5. This point is formulated as a general transaction-cost problem by Babbage:

> The cost to the purchaser, is the price he pays for any article, added to the cost of verifying the fact of its having that degree of goodness for which he contracts. In some cases the goodness of the article is evident on mere inspection; and in those cases there is not much difference of price at different shops. The goodness of loaf sugar, for instance, can be discerned almost at glance; and the consequence is, that the price is so uniform,

and the profit so small, that no grocer is at all anxious to sell it; whilst, on the other hand, tea, of which it is exceedingly difficult to judge, and which can be adulterated by mixture so as to deceive the skill of even the practised eye, has a great variety of different prices, and is that article which every grocer is most anxious to sell to his customers. The difficulty and expense of verification are in some instances so great as to justify the deviation from well-established principles. Thus it is a general maxim that Government can purchase any article at a cheaper rate than that at which they can manufacture it themselves. But it has, nevertheless, been considered more economical to build extensive flour-mills, (such as those at Deptford,) and to grind their own corn, than to verify each sack of purchased flour, and to employ persons in devising methods of detecting the new modes of adulteration which might be continually resorted to. (cit. after Mill, 1848, 133)

Thus we seem in commodity production to have a generalised version of the law of Gresham: the poor quality drives the good quality out, unless some countervailing measures are made. There seem to be two major models: in the case of many producers of a standardised commodity you have to enforce the standards (by the invention of measurement instruments and control institutions). The other possibility is to introduce 'monopolistic competition' where each producer is producing a special variant of the good for which he is solely responsible and is risking to injure both sales and goodwill if standards are not high and stable. However, the differentiation of commodities through monopolistic competition has to take place within the confines of an overall definition of commodities unless the informational burden upon the users shall explode and the possible sales volume shall consequently diminish radically. In the information-rich world described above, the buyers must normally be able to assume that their accumulated knowledge of the characteristics of the commodity-type will more or less hold for the next instance of it. This assumption is of general necessity, but in some cases of inter-industrial deliveries it is of critical importance for the quality of the product in which it will be embodied.

This assumption can be formulated generally as a principle of commodity abstraction. It is not only found in economic practice but is also fundamental in engineering design where an attempt is made to construct clear-cut interfaces between different parts of a system. If this principle is followed it is normally possible to concentrate attention on a single part of the system. The possibilities of faults and problems originating in the environment can only be transferred through the interface, and changes in a specific part can be made locally as long as the changes do not influence the characteristics of the part/product defined by the interface. The fixed definition of a commodity gives such an interface between producers and users. In this way the transaction costs are lowered because less information needs to be transferred. The costs of future innovation are also lowered *if* changes in processes of the producer or the user can take place without challenging the given interface.

The basis of full-fledged commodity abstraction is, of course, an artefact which is reproducible. The principle of recurrence comes before the principle of commodity abstraction. However, the abstraction principle says something more since two instances/implementations of a given commodity type/species will never be *totally* equal. Therefore, some abstraction from real differences between instances of a commodity is involved in 'commodity abstraction'. This abstraction is based on the hypothesis that the differences between the 'idea' and 'specification' of the commodity type and its implementation are of minor importance. When making abstractions about nature we have to accept the real world, but in the design of artefacts like computer programs or economic commodities, we may try to mould these items in a way which makes them especially fit for later abstractions. In this way a name together with a few parameters is sufficient to describe (or command) fully an object or an action. However, in reality the full subsumption of complex computer programs or economic systems under the imperative of abstraction is but a dream. The never-ending fight against complexity and for better possibilities of abstraction is the central theme in the development of computer science and this has been of much use in the design of computers and programming languages. In economic theory there has been a fight against the complexity of the *models* of the techno-economic system and it has even been possible to create theoretical models which are only seemingly complex. However, most of this work has been without relevance for the fight against the real complexity of the techno-economic system.

The fight against real complexity and the related impossible amounts of information needed for acting economically has developed in a piecemeal fashion based on the 'local' environment of the actor. The hypothesis that the characteristics of a commodity are reflected by a few abstracted notions is based on earlier trials and errors of the user or his references. But this process will only lead to relatively stable and usable knowledge provided that there is a high degree of stability of the producer, a stability which is enforced by the producer's organisation as well as through many inter-organisational customs and institutions. This institutional framework was partly established to cope with earlier violations of the principle of commodity abstraction. The user will not only demand a commodity but also a supplier-behaviour according to this principle. But as soon as the interface between the producer and the user is organised according to the principle it becomes extremely difficult to make changes. First of all, routines on both sides of the interface are developed under the assumption of a stable interface and there is little hope of an agreement on a strategy of change. Furthermore, the information flows are no longer sufficient for telling about possible reactions to changes in the interface. Finally, the amount of information needed if the principle of commodity abstraction is broken may become quite uncertain.

For such diverse reasons the parties in commodity exchange tend to become the servants or slaves of an interface between them (a 'dominant design') which was originally one of many design alternatives. The near-classical example is the QWERTY keyboard of typewriters (Arthur, 1983; David, 1985/93) which reflects the well-established standard for the sequence of keys slowing down the typing speed by 20-40% *vis-à-vis* the Dvorak keyboard.[19] This standard was developed and frozen in the years around 1870, especially by the mechanics of Remington and Sons which extended production from arms to typewriters. A major design criterion was to avoid the mechanical type-bars clashing and jamming when struck rapidly after each other. This and other problems of the original design were, however, more or less solved in a few years, but the interface between the typist and the machine (and part of the interface between the producer and the user) was already fixed and has remained so to this day. All the rivalling (and better) designs which were invented in the following years (including Dvorak's solution) were beaten by, e.g., the economics of scale of Remington and the reluctance to skip learning effects by the first few typists (and the technical interrelatedness to the firms producing typing courses). In this way a quasi-irreversible path of maintenance and evolution was created. Today, the problem is certainly not a question of technical inflexibility. But nevertheless we are all (including the writer of this book) facing the lock-in to a design and a path chosen around 1870.

The QWERTY keyboard does not include a full specification of the typewriter. A few other facts of its functionality seen from the viewpoint of the user are also part of the specification (including the use of typewriter ribbon, etc.). But the design is seldom defined fully in functional terms. The user might conceive information on the materials used (iron etc.) as a proxy for its usability characteristics. These explicit characteristics define a paradigm according to which the concrete instances of typewriters and computer keyboards are produced and bought without too much difficulty of communication. But they clearly abstract from many of the (in principle) infinitely many characteristics of each unique instance of typewriters. A few of these might however be part of a parameterised interface which allows the user to determine easily on major issues (correctability of written text). When some authors (Sahal, 1985; Saviotti, 1988) speak of a characteristics approach they are discussing such parameterised interfaces (as size, horsepower and energy-use of transport equipment; memory and clock speed of computers). Both these attempts and the characteristics approach in general presuppose a highly structured and stable world organised according to the principle of commodity abstraction where you can consider the vast majority of the infinitely many possible characteristics of an instance of a commodity to be either OK or irrelevant.

2.6.2. The principle of interactive learning in product innovation

In a techno-economic system organised according to the principle of commodity abstraction there is little room for innovation. The reason is that innovations will not take the finished and information-poor form of the time when they are first introduced on the market. In such and similar cases the principle of commodity abstraction cannot be applied directly. It is more appropriate to consider new or easily adaptable products as 'proto-commodities' which through a process of 'commodification' governed by Schumpeterian entrepreneurs (or by a more gradual evolutionary process) may be transformed into real 'commodities' (in my sense). When a new proto-commodity implies some kind of jump, it has to be constructed by means of anything you can lay your hands on, according to what in Danish is called 'the principle of the available nails' or in French the principle of *bricolage*.[20] This principle does not lead to well-defined and stable solutions but to constructs which require a complex process of debugging. This process takes the form of a development path which may be heavily dependent on some idiosyncrasies of the original construct and its more or less randomly chosen components.

The idea of innovation as the construction of a proto-commodity[21] by whatever means available comes close to Schumpeter's (1912/34, 65 f.) definition of innovation:

To produce means to combine materials and forces within our reach. ... To produce other things, or the same things by a different method, means to combine these materials and forces differently. In so far as the 'new combination' may in time grow out of the old by continuous adjustment in small steps, there is certainly change, possibly growth, but neither a new phenomenon nor development in our sense. In so far as this is not the case, and the new combinations appear discontinuously, then the phenomenon characterizing [Schumpeterian] development emerges. For reasons of expository convenience, henceforth, we shall only mean the latter when we speak of [innovations or] new combinations of productive means. Development in our sense is then defined by the carrying out of new combinations.'

What is lacking in this definition (and in Schumpeter's subsequent discussions) is an explicit analysis of how the unfinished character of the new product (or process) is dealt with. The reason is that Schumpeter is not interested in the process of invention.[22] By emphasising the introduction of full-fledged commodities, he avoids the reintroduction of the problem of invention in the post-innovation phase. At the same time he has to blame all difficulties of and all resistance to innovation on established firms and on non-inventive consumers. This judgement changes when we consider the innovation to be a proto-commodity which is still unfinished when compared with the basic design principle of the components of complex systems. From this viewpoint the 'new combination' has to undergo a sustained process of debugging before it can be simple enough in terms of informational requirements to be of relevance for generalised use.

This debugging clearly requires another type of information flow than cases where the principle of commodity abstraction can be presupposed. The principle underlying the gradual commodification of the proto-commodity may be called 'the principle of interactive learning' in product innovation and product modification. This principle can be seen as a generalisation of Arrow's (1962/83) analysis of 'learning by doing' to deal with the problem of product innovation. In this case learning by doing is really 'learning by using' (Rosenberg, 1982, ch. 6) new products (which I will assume to be machines or other producers' goods). One could also speak of a kind of knowledge accumulation founded in the trial and error in the use of the product. In the case of computer programs such knowledge accumulation will often take the form of a list of faults and major problems ('bugs') and another list of new facilities which may be important to the user. This learning and knowledge accumulation is (to a large extent) located in the using sector and some of its results can be implemented without changing the machinery or the programs (disembodied technical change). Other results concern the improvement of the machinery. If we presuppose a strict division of labour, such ideas cannot be implemented without a product development in the machine-producing sector. This sector is, furthermore, the main supplier of knowledge about the technological possibilities for such improvements.

Under such (plausible) assumptions we have learning by doing/using in the machine-using sector but some of the possible results of this learning can only be realised by the machine-producing sector. Therefore, successful interaction between the two sectors is a precondition for much of the technical change embodied in improved machinery.

Even if I have developed this argument in terms of the production, improvement and use of capital goods, it seems obvious that similar situations are pervasive. Whenever there is a vertical division of labour, some of the problems discovered by the machinery case must be of relevance. This result can be called the principle of interactive learning in product innovation. In order to function, the interface between the two parties cannot be fully frozen and we will find linkage-effects which are not fully accounted for by the economic exchanges of commodities and money, not even if we include commodified information. We always find a two-way exchange of the results of learning processes in relation to product development and product use, which clearly breaks with the principle of commodity abstraction.

This way of looking at inter-firm relations is to some extent developed by Arrow himself:

... the whole idea of a firm with definite boundaries cannot be maintained intact. For example, the customers of a firm are, to some extent, part of it ... There are direct information flows from customers in the form of complaints, requests for product alteration or special services, or threats to change to another firm, in addition to the anonymous alterations of demand at a given price which constitute the sole information link between a firm and its market in neoclassical theory. Some employees of a firm will have closer links to customers than to at least some of the other employees. (Arrow, 1973/83, 147)

If we take a single producer-user pair in capital goods, there will be employees with a double responsibility for the product development and use of a specific type of machine. The personnel of the user of a chemical plant (or a complex computer program) will gradually find major faults and 'bugs' and even some solutions to them. To the extent such results are available to the machinery producer they form part of the 'agenda' for future product innovations. At the same time the developers of the machinery producer influence (through price and quality) the production process and maybe even the possibilities of product development of the machine user. Between the two parties there is an informal but contract-like relationship: there is an expectation and probably a demand for a certain degree of 'reciprocity as a condition for continuing with the relationship.' (Nelson, 1987, 80)

Not all producer-user relations are equally important for product development. Some users play an active role and presumably the resulting new products are well-suited for them. Other users reap the benefits from improved goods without playing an active (and costly) role in the innovation process. Still others have to accept the standardised goods on the market even if they are not well-suited for their needs. The sequence of events may have started when learning by one firm was transferred to the supplier of a capital good, but now the results of the original learning are embodied in the equipment purchased by the late-coming user, irrespective of his specific situation. He has to accept the principle of commodity abstraction.

2.7. Conclusions on paradigms

Commodity abstraction and interactive learning appear to reflect two opposing principles for designing the producer-user interface. But this is not necessarily the case. Even within the limits of the QWERTY interface, there are plenty of possibilities for gradual improvements of the typewriter beyond the reach of the initial entrepreneurial act of Remington and others. As formulated above, interactive learning has more to do with the intra-paradigmatic change than with the establishment of new paradigms. To turn to the latter issue we should note that each of the interfaces is defined by two parties and is therefore not freely variable, at least not in the short- and medium-term. In this way each interface represents a kind of objective or inter-subjective knowledge (Popper, 1972) which creates a certain degree of predictability and which has to be taken into account unless the firm is to be highly vulnerable. In some areas this objective knowledge has the character of economic paradigms, in other areas this knowledge tells determined Schumpeterian entrepreneurs that there are possibilities of establishing new economic paradigms.

A Schumpeterian manager who operates in an economic world without full commodity abstraction has to relate to a large number of actual and potential customers or suppliers from whom information is gathered. In each case an 'information channel' may be established and the treatment of the resulting information within the innovating firm analysed (Arrow,

1974). In this context it is important to note (as a consequence of the behavioural analysis) that only a tiny part of the potential relationships between different firms can be developed into well-established information channels. And only a small part of the actual channels are central to incremental innovation. It is such channels which provide a good deal of the information used in technological development. Any other possibility would by far transcend the information-processing capabilities of firms, even the innovative firms. A major task of the formulation of a strategy of change concerns the pointing out and the evaluation of the possible relative importance of such channels. It also implies solutions to the problem of how the information is transformed into items on the agenda of technological development of the firm. Such tasks of the firm do not exist in the Walrasian version of neoclassical economics. Here all the potential relationships of a firm are checked each time it makes a decision:

In classical maximizing theory, it is implicit that the values of all relevant variables are at all moments under consideration. All variables are therefore *agenda* of the organization, that is, their values have always to be chosen. On the other hand, it is a commonplace of everyday observation and of studies of organization that the difficulty of arranging that a potential decision variable be recognized as such may be much greater than that of choosing a value for it. (Arrow, 1974, 47)

For this reason organisations normally keep most variables (including the character of most of their output and input interfaces) away from their agenda. However, some of the information from selected customers (and suppliers) is allowed more or less automatically to change the agenda of the firm. In this way the organisation of a machine producer is, e.g., able to deal with minor proposals for product change which come through the information channels to selected customers. Such channels may also help to change the agenda of major product developments, i.e., to create an organisational awareness of the need for a real shift in product characteristics. This may be crucial since, according to Arrow (1974, 47), 'innovation by firms is in many cases simply a question of putting an item on its agenda before other firms do.'

The degree of standardisation of the different interfaces decides whether the related information channels are relevant for changing the agenda of product and process development. Some of the studies of the product life cycle[23] may be reinterpreted in this perspective to show how and why the innovative importance of information channels between producers and users is changing over time. This discussion will be developed at the industry level even if we have the problems of evolutionary strategies of individual firms in mind. However, it should be noted that such an approach tends to wipe out the distinction between Schumpeterian managers and entrepreneurs. The industry-oriented product life cycle starts with the introduction of a new product on the market; this may be considered as an act by one type of Schumpeterian entrepreneur. However, the discussion tends to concentrate on the cluster of still more minor innovations which follow this primary innovation. The first of these changes may be considered as the act of another type of Schumpeterian entrepreneur. In the phase immediately following the original innovation the establishment of a market for the product is a major task, and the innovation-agenda of the firms of the industry focuses on product characteristics. For this reason firms with well-established information channels to sophisticated customers have a comparative advantage in the creation of (minor) innovations. However, many new users have neither the possibility nor the motivation to understand all the considerations behind the design of the product/good. To them the product/good tends to become an abstraction, a given thing with well-defined attributes which may be represented by a name, a price and some extra information delivered by the seller in a product specification which is part of the sales contract. In other words the product/good tends to become a commodity

In this situation there is a possibility of establishing a dominant product design or an economic paradigm which implies a standardisation of the interface *vis-à-vis* the users and a simplification of the related information channels. At the same time entrepreneurial interest can be concentrated on rationalising the process of production. This final solution to the confusing manifold of design alternatives may be illustrated in relation to the automobile industry. The securing of the dominance of a specific design may be considered as the act of a third type of entrepreneur. The victory is not due to intrinsic user characteristics of the product but simply that it is redesigned in a way that makes it well-suited for large-scale production and that this large scale is implemented with a right timing when the market is appearing and before others competitors are ready for this dramatic shift in the competitive game. The shift came, of course, primarily with Ford T and its productive capacity which were running in forward of the demand. With this change the innovation-agenda immediately shifts towards the problems of mass production and mass consumption. The information channels to the suppliers are strengthened, and interactive learning focuses on process technology and helps the development of (minor) process innovations during the growth phase of the product life cycle. But the importance of a more abstract approach *vis-à-vis* the consumers should not be underestimated. This may explain why Schumpeter uses the formulation that 'new commodities or new qualities *or new quantities* of commodities are forced upon the public by the initiative of entrepreneurs' (Schumpeter, 1928/51, 65). The spread of Ford T to millions of buyers is an innovation which cannot be reduced to mass production and cheaper cars but also to efforts to persuade, teach and provide service to the customers. After a period a process paradigm is emerging, defining precisely the interface *vis-à-vis* the suppliers. From that point the industry enters a period of 'maturity' where innovation is primarily cost reducing and takes place in the supplier industries with relation to the general technological development of the economy but with relatively little relation to the users. The industry is at this point facing rapidly decreasing informational barriers to entry. The acts of Schumpeterian entrepreneurship are a history of the past—in this particular industry and for the time being.

When looking back on the reinterpretation of the industry-oriented theory of product life cycles it is clear that it tends to make management and entrepreneurship a function of the stage of commodification. Real managers and entrepreneurs are probably facing possibilities of sticking to a flexible state and reversing the life cycle from an inflexible to a flexible state. In this situation it is important to note that there are two faces of an innovation-oriented information channel and an organisational agenda for innovation. In an early period channels and agendas are established with a clear understanding of their purpose. But gradually they are facing a 'productivity dilemma' (Abernathy, 1978). There is much investment of human and organisational capital which is dependent on the specific set-up. It often becomes a purpose in itself to use the given framework, to formalise informal agreements and to decrease their costs. When a major shift in innovative possibilities occurs, the given structure is very seldom prepared. This fact was taken very seriously by Schumpeter (1912/34) who connected (major) innovations with new men and new firms. In well-established but innovative firms (which Schumpeter later (1942/87) became aware of) change may be supported by the circulation of elites (Arrow, 1974, 59). However, in the case of well-established producer-user relationships changes may be even more difficult since the two parties involved have to make a coordinated move. The chance of avoiding such fixed inter-organisational relationship may be a major reason why radical innovation is often combined with 'new men and new firms'.

Another dilemma is founded in the conflicting interests of the parties related to a changing interface. Thus the change from proto-commodity to full commodity is (normally) not in the interest of the first sophisticated users. They reap the benefits of tailor-suited equipment while commodification may mean less care about their specific requirements and at the same time spread of the results of learning by using to competing firms. This may explain some of the controversy over the relative importance of users and producers in innovation. The gradualist approach to innovation has studied the emergence of proto-commodities in close interaction between producers, and often the initiative is taken by the users. This has to some extent become the dominant approach (see the review by Mowery and Rosenberg, 1979/82) because the creation of proto-commodities is the main task of product development. But in the transformation of proto-commodities into full commodities (in my sense) the definition of the roles changes. Now the producer is the 'king' who is 'forcing upon the public, a new commodity' (Schumpeter, 1928/51, 64). In this task it is (Schumpeter, 1912/34, 65) 'the producer who as a rule initiates economic change, and consumers who are educated by him if necessary; they are, as it were, taught to want new things, or things which differ in some respect from those which they have been in the habit of using.'

This viewpoint presupposes 'debugged' and otherwise full-fledged commodities since we would otherwise have to consider feed-back from the consumers. A product innovation might in some cases take this form from the very beginning (simple goods, often for consumers' markets) but the modern analysis of innovation and evolution has shown that this is the exception rather than the rule. In the normal case the 'thing' introduced on the market is a proto-commodity which only after a period of trial and error is stabilised into a product-commodity with a higher degree of correspondence to the principle of commodity abstraction. When the market is saturated and the system can be thought to be in some kind of equilibrium, we see yet another inversion in the roles of producers and users. In this state (Schumpeter, 1912/34, 65) ,'it is permissible and even necessary to consider customers' wants as an independent and indeed the fundamental force'. In other words, producers and users share the power of creating proto-commodities, the producer is dominant in much of the commodification and diffusion and the 'consumer is king' (Samuelson) in the state of the 'circular flow'.

Notes

1 It is not easy to select such standard cases for evolutionary analysis. There are many types of evolutionary analysis and each of them will prefer cases which are well-suited to its special problems and possibilities. Tradition may, however, help us to agree upon some standard areas of a give and take between theoretical analysis and 'detailed historical case studies' (Schumpeter, 1949/51a, 311). In my opinion, Schumpeter suggests such a standard area rather than a simple testing of his own kind of approach when he points to the process of 'railroadization' (Schumpeter, 1939, 304). He is actually relating to a permanent tradition of applying the case of railways and their economic repercussions. Actually, the pricing of railway services and the overall process of railroadization are central themes from the early partial-equilibrium and evolutionary economic analysis (Ekelund and Hébert, 1975/90, 303 ff., 315, 372 486 f., 510, 602, 607) via Marshall (1890/61; 1919) and Schumpeter (1939) to the new economic history (O'Brien, 1977) and Chandler (1977, part II; McCraw, 1988).
2 I have chosen to stick to Schumpeter's American English coining of the term of 'railroadization' to emphasise that it is used in Schumpeter's broad sense. It is not easy to transform it into a well-defined technical term.
3 The major alternative as a paradigmatic example is what may be called 'automobilization'. Thus Schumpeter remarks: 'The automobile industry affords a good example of a purely entrepreneurial achievement turning to new uses not only existing resources but also existing technology'. (Schumpeter, 1939, 415) 'The automobile industry led in every upswing and out of every downswing throughout the period [1919-1929], in fact far beyond it, and continued in the Kondratieff recession [after 1929] to qualify as well for the role of standard example for the processes embodied in our model as it had done in the upswing.' (Schumpeter, 1939, 772). Some developments of this case are found in *Business Cycles* (e.g., Schumpeter, 1939, 415-418, 772-777). Still, this is not the example of the young

4   Which even is a core case of 'the process of capitalist evolution—economic evolution as conditioning, and being conditioned by, the institutional pattern of bourgeois society' (Schumpeter, 1939, 304).
5   Schumpeter never made a systematic development of the paradigm of railroadization in relation to his basic evolutionary ideas. However, much material is presented in the section 'Railroadization' in *Business Cycles* (1939, 325-351) and in other parts of this book (e.g., 17, 73, 101, 113, 146, 158 f., 167 f., 277 f., 280, 291 f., 303 f., 314, 319 f., 352, 354, 357, 359, 361-363, 366, 368 f., 377, 385, 395 f., 402-408, 413 f., 488, 523, 569, 630, 632, 663, 686, 760, 878, 884). That this is not a particular interest of the elderly Schumpeter is revealed by many shorter references to the example in his first book (1908, 189 f., 245, 247, 250, 252 f., 451 (?), 512-514, 568, 584, 608-612). Further references are scattered throughout his works (e.g., 1942/87, 68, 83, 99, 119, 132; 1912/34, 62, 66, 84, 215; see also 1912/26, 93-95, 123, 321).
6   The task of the researcher is to give a theoretical explanation of the observed fitness, i.e., to define theoretical fitness independently of the observed fitness. See below.
7   Both in the construction of commodity group j and commodity group i we should take into account that 'composites derive additional justification if their constituents are, owing to their place in the economic organism, all exposed to similar external and internal influences. As with group prices, it is necessary to bear in mind that composites may seriously obscure what precisely is the essential fact about the cyclical process of economic evolution.' (Schumpeter, 1939, 484)
8   Or 'equivalence classes' (Winter, 1987). However, all the elements of a formal equivalence class are, of course, formally equal but in evolutionary analysis most 'classes' of routines are considered to have an important degree of heterogeneity which may change over time. This creates several difficulties which are somehow related to the discussion on the paradoxes of mathematical set theory. One partial solution is to specify sharply which types of elements are members of the set under discussion.
9   Since the argument is operating in terms of frequencies of classes of routines there is still the possibility that one could find evolution by changing the level of aggregation of the analysis. Maybe some kinds of change in the composition of the classes of routines will not influence the relative frequencies under study. Therefore, we should make explicit that we are talking of evolution which is discernible at the given level of aggregation.
10 But here we have the problem of eliminating changes in relative frequencies due to oscillations in the importance of the capital goods industries.
11 Witt (1985) has a similar distinction between coordinating and decoordinating tendencies which shape the movements within the price-quantity-quality space.
12 It may sometimes be convenient to extend the range of operations and allow the α-operator to take a Δ-state and return a Δ-state while the β-operator may take a Σ-state and return a Σ-state.
13 The text of the paper is partly included in *The Theory of Economic Development* and this may be the reason why it has not been used in any of the collections of Schumpeter's papers.
14 After many controversies Schumpeter left his original concepts: 'It would, perhaps, be best to drop the terms statics and dynamics altogether. Certainly they are misnomers, when used in the sense given to them in the text, and care should be taken not to think of them by way of analogy with their meanings in mechanics' (Schumpeter, 1928/51, 60).
15 A further development of the arguments on economic paradigms (sections 2.5-2.7) is found in Andersen (1991a).
16 Roughly we may say that both standard economics and its alternatives have moved away from rather than towards a standpoint which makes an understanding of

Schumpeter easy. Let us concentrate on the question of a new combination in terms of using the factors of production to supply a new type of commodity to consumers. In relation to the Walrasian system we may think in terms of a set of commodity types which has n elements which in turn defines n equivalence classes of products and n abstract industries which produce these products. A huge (supposedly infinite) number of Walrasian entrepreneurs are organising the production for each commodity type. Moreover, the set of commodity types at time t and $t + 1$ is the same. In this setting Schumpeter simply says that the number of elements in the set of commodity types at time $t + 1$ is increased to $n + 1$ by an innovation. The idea is apparently simple and it is even well-founded in the economic analysis in the pre-Walrasian period. This background may be revealed by a study of the concept of 'commodities' and related concepts of 'products' and 'goods' have been used in the different traditions of economic analysis. (The usage of the concept is discussed by Milgate, 1987.) As far as I understand, the commodity concept was originally a generalisation based on primary staple goods which were major items of trade during the 17th and 18th centuries. From the usage in this area, the concept went into theoretical economics, but, unfortunately, with many different meanings. For instance, we find in Ricardo's *Principles* a discussion of reproducible commodities *versus* unique commodities. The present analysis is related to the former meaning but the double meaning opened a door for much confusion. In the present context we may say that it is only the former meaning which is bound to institutional attributes like a high degree of division of labour, private ownership, and a consequent need of standardising the characteristics of the commodity. Therefore, there is an essential socio-economic distinction between 'commodities' in the narrow sense and 'products' or 'goods' which have no such institutional attributes. The main trend in economics has been to abandon the commodity concept for this very reason (or to use it as a synonym of 'good'), and the change started in the English literature with Marshall's *Principles of Economics*. Instead the more general concept 'good'— which includes usable things which are non-produced and even non-traded—were used. However, the concept of 'good' is so general that it has no direct relation to the evolution of the structure of the economy, even if economists continue to speak about industries producing different goods under the condition of specific production functions and about markets for these goods. These distinctions are only possible if institutional factors are taken into consideration. Without this socio-economic definition of goods (i.e., commodities) the techno-economic system falls apart into its basic components: firms and households. This point was made clear by discussions following the theories of monopolistic/imperfect competition implying that each firm produces goods which are in some degree special to itself (Shackle, 1967, 65 ff.). If was not least Chamberlin (who in other respects has some affinity to Schumpeter) who helped to dissolve the idea of commodity types and industries (Chamberlin, 1933/62; Chamberlin, 1953). The conclusion tends to be to leave the question of innovation and non-innovation to the subjective judgement of individual firms (and households). It is, however, difficult to see how evolution is to take place in such an unstructured context. It is more easy to develop in a context where some sort of commodity concept is rescued by means of the idea of economic paradigms.

17 There is no necessary incompatibility between the two approaches, and Simon (1981, 41 f.) quotes approvingly central passages of Hayek's (1948) views on the 'division of knowledge' in complex economic systems. But still we have two relatively independent analytical traditions.

18 To clarify my point I propose a modified version of Dosi's (1982, 148) definitions of technological paradigms and technological trajectories and I emphasise this point by talking of economic paradigms. I have sharpened the differences *vis-à-vis* Dosi in order to clarify the area taken up in the present argument.

19 Here I take Arthur's and David's story for granted. Some doubts about the superiority of the QWERTY layout has been raised by Liebowitz and Margolis (1990).

20 See Jacob (1981/85, ch. 2; partly published as an article in *Science*) The jumps of biological evolution are in Jacob's eyes analogues to the unworkmanlike tinkering of

a *bricoleur* which he contrasts to engineering principles of functional design. He argues that the difference can be seen in the fact that *bricolage* leads to a set of different approaches to the same function while the solutions of engineers (like cars, cameras and ball-pens) are all closely similar to each other. But Jacob is here looking at areas where a dominant design has appeared. A closer look at earlier stages will show that technological development also jumps according to the principle of the available nails.

21 Or a new production process which does not allow full process abstraction and which, consequently, gives big problems for price-setting.
22 More fundamental problems relating to Schumpeter's lack of recognition of the role of novelty are emphasised by Witt (1992b).
23 Much of their development of the argument is found in Abernathy (1978), Abernathy and Utterback (1975; 1978), Abernathy and Townsend (1975), Abernathy and Clark (1985), and Clark (1985).

3. Density-dependent diffusion and innovation

3.1. The network of interrelations

3.1.1. Integrating 'ecological' and evolutionary analysis

Many of the advances of the new evolutionary economics have been made by making simplifying assumptions that Schumpeter was not willing to make. Many of these assumptions seem to lead away from the evolutionary perspective but some of them have, nevertheless, clarified important aspects of evolutionary processes. Of special importance is the idea to stop introducing new variety into our evolutionary models. We thus have a given set of production routines (commodity types, production methods, etc.) which firms can apply. The frequency of the application of a certain routine will increase because of the supernormal growth rate of the firms applying this routine or because of the supernormal propensity to imitate firms applying this routine. Other routines will become less and less frequent. However, in a longer time perspective things may be much more complicated because of the density-dependence of the growth rates of the routines. A typical pattern is that the growth rate of the frequency of a routine decreases as it is applied more often. Another pattern is that the density of the application of one routine influences the growth rate of another routine. Some routines help each other become more frequent while other routines compete with each other. The broader environment in which they function influences the performance of them all. Such patterns are the subject of classical ecological analysis, both as a biological discipline and as a metaphorical expression which may be applied to practically all areas of study.

The 'ecological' part of the analysis of economic evolution has to some degree been present in the last chapter but little explicit attention has been given to it. One reason why it has been neglected is that while all kinds of economics have some 'ecological' components, evolutionary economics analysis is special because it makes a broader synthesis which includes 'ecological' aspects. But Schumpeter studied the process of economic evolution from the viewpoint of the innovating entrepreneur and this meant that he tended to have too much evolution and too little 'ecology', too much analysis of variety-creation and too little analysis of the economic structure. Several of the economists whom he inspired have tried to remedy this situation. From Dahmén (1950/70), Perroux (1955/69), and onward, a wish to combine an evolutionary and a structuralist perspective has existed, but relatively few convincing results have emerged. The reason for this lack of success may partly be that we are facing a basic methodological difference between the study of innovation and the study of diffusion. This distinction has been emphasised by Witt (1993b) who insists on the fact that innovation, by definition, involves emerging properties which cannot be known *ex ante*. This problem makes it important to distinguish sharply 'prerevelation analysis' and 'postrevelation analysis'. At present we notice that it appears to be helpful for the development of evolutionary economics to begin with studies where variety-creation is (partially) blocked in order to focus on the problems of (simpler forms of) interdependence. The tools for this kind of analysis have mainly been developed within biology, but it must be emphasised that these tools are very general and that pioneers like Lotka and Volterra were many-sided scientists who were also interested in economics (Kingsland, 1985; and the relevant articles in Eatwell *et al.*, 1987).

'Ecological' analysis is not an independent area in the development of new evolutionary economics. Nevertheless, certain approaches may emphasise 'ecological' starting points for the analysis. This gives a special flavour to their contributions to evolutionary economics, which synthesises creation of new variety and its diffusion into wide-spread economic practice. To be more specific, such 'ecological' approaches take their point of departure in the diffusion of innovations rather than in creation of innovations. They may often be recognised by the 'trade mark' of diffusion studies as well as ecological studies, namely the S-curve or, more specifically, the logistic curve of diffusion. Such a pattern appears to occur empirically in most diffusion studies (see Stoneman, 1983; Rogers, 1962/83) but there is a danger of overdoing attempts at 'curve fitting'. From the very first modern attempts to find examples of the logistic curve in population studies and economic life (Pearl and Reed, 1920; Kuznets, 1930/67), there have been surprisingly sharp critiques of this practice (Kingsland, 1985).[1] However, for present purposes it is sufficient to note that there are plenty of examples from economic history where the diffusion of a successful innovation can be rather well-represented by a logistic curve. This is not to say that we cannot find other mathematical representations of the facts, but this particular type of an S-shaped curve reflects the simplest mathematical model which should be seen as a starting point for theoretical studies as well as for sharper hypotheses about the facts. Furthermore, we shall see that it is the most direct way to study competition and between the application of different routines.

There is a trade-off between on the one hand, the emphasis on the complexity of the 'ecological' interactions in the economic system and on the other hand, the emphasis on the evolutionary processes within this system. Since we are here biased towards the latter option, we have to be relatively crude in our structuring of the system. In this respect, the situation is similar to that of evolutionary ecology within biology, which tries to combine both areas of study. However, in this case a strong tradition of non-evolutionary ecology has gradually been expanded by relatively successful attempts to overcome the split between evolutionary studies and studies of interacting populations with given characteristics (Roughgarden, 1979; Pianka, 1983). In the social sciences we have more proposals than actual results. One series of relatively ambitious proposals about combining 'ecology' and evolution in the study of social and economic affairs has been put forward by Boulding (1978; 1981). Somewhat more concrete approaches have been developed by Hannan and Freeman (1989) as well as by Cavalli-Sforza and Feldman (1981) and by Boyd and Richerson (1985). Still, the most important approaches are found in a series of studies of the diffusion process by economists who try to develop an evolutionary economics. For example, we may notice the modelling work by Iwai (1984/93; 1984) and Silverberg et al. (1988) as well as the surveys by Metcalfe (1988) and Silverberg (1988). However, none of these latter studies are explicitly developing the relationship with 'ecological' modes of thinking which will be the centre of the present exposition.[2]

3.1.2. The unfinished growth pole analysis as an example

As a motivation for the relatively formal kind of analysis which will be developed in this chapter, we shall start with an example of the uneasy relationship, or mismarriage, between evolutionary and structuralist approaches. The example is the growth pole theory which was developed by Perroux (1955/69; other papers in 1969). This theory is a member of a family of theories about the transformation of the economic system which derived their main inspiration from Schumpeter. Other examples are Dahmén's (1950/70; 1988) theory of 'development blocks' and Hirschman's (1958; 1987) theory of backward and forward 'linkages'. The idea behind the growth pole theory appears to be taken from Schumpeter's discussion of direct and indirect effects of a radical innovation (e.g. in Schumpeter, 1928/51, part II). A major innovation may be seen as creating the basis for a whole series of more or less adaptive decisions during a shorter or longer epoch of time. Some of these are performed within existing routines while others are innovative, but supposedly of an adaptive and incremental character ('clusters of innovation). Such sequences of decisions may be captured by the notion of growth poles in the industrial system consisting of propellant industries (the primary, 'autonomous' innovators) and impelled industries (showing adaptive response, including 'induced' innovation). The development power of a region or nation is to some extent dependent on the existence of propellant industries within its realm and many national policies can be understood as attempting to monopolise the resultant forces of development. Even if Perroux criticises the national monopolisation of the inducements to develop in some of his works, he is clearly sketching a production and linkage approach to regional and national systems of innovation (see Lundvall, 1992).

However, this set of ideas was not clear to all his followers (see Brookfield, 1975, 105 ff.). Many of them were not interested in the evolutionary 'laws of succession' but rather in the 'laws of coexistence' between the different parts of the industrial system and ultimately phrased their ideas in terms of Leontief's input-output analysis rather than in Schumpeterian terms. In their transformation of the idea of growth poles from one paradigmatic context to a radically different one, they could use the fact that Perroux was developing his ideas in terms of industries and that he himself suggested applying input-output analytic tools to the phenomenon of growth poles. They translated the theory into an input–output language with (temporarily) fixed technical coefficients in the following way: large technical coefficients can be taken as proxies for important 'linkages' (Hirschman, 1987) or propelling forces; 'industrial complexes' are parts of the industrial system connected by strong 'linkages'; the cores of the 'industrial complexes' can partly be found by means of the inverted input-output matrix which shows the direct and indirect inputs used for one unit of output of each industry of the industrial system. The policy prescription for development policy was then to invest in important core industries which for one reason or another were not present in the nation under consideration. The rest of the industrial system would then be constructed or renewed by means of the propellant forces from the core industry which was also the core in a system of innovation. However, the translation of Perroux's argument is, unfortunately, radically wrong. The tight 'linking' of industries revealed by the input-output tables of the most advanced countries has no necessary connection to growth poles. On the contrary, it probably indicates a 'mature' situation with routine deliveries where there are few possibilities of, and little impetus to, change and development. The 'industrial complex' which is found in this way may be like a mature ecosystem where all 'niches' are occupied while a growth pole is like an immature ecosystem which is trying to establish itself through rapid evolution. Unfortunately, the latter situation is nearly impossible to catch by means of input-output analysis. The rough tools of absorbing Markov chains appear to be more adequate but, to my knowledge, they were not applied in growth pole analysis.

The sketched formalisation of growth pole analysis led to a boom in its application in regional and development economics. But the success was quite short-term because the studies had little to do with change and innovative investment strategies and much more with the interdependence of a well-established industrial system. This led to a neglect of Perroux's concentration on disequilibrium and on the character of the exploration of the space of investment possibilities which constitute the foundations for the whole approach.[3] Furthermore, great problems were created because of the too-ready translation of the search space connected to economic decision-making into a space dominated by geographical or national distances, a translation which clearly neglected Perroux's (1950/69) original warnings. The boom ended (see Brookfield, 1975, 92 f.) in 'immense confusion', mainly because of a neglect of the fact that 'the activity creating a growth pole was essentially a sectoral and geographical disturbance not because of its larger than average size, nor because of its higher multiplier, but because it was [... a radical] *innovation.*'

An important conclusion of the story of growth pole analysis appears to be the necessity of making very explicit the basic assumptions. The wide-spread non-evolutionary modes of thinking and inappropriate analytical tools may otherwise provide a propensity to drift away from the assumptions and end up in confusion and even nonsensical statements. To avoid this it is important to emphasise the assumptions concerning the aspect of discontinuity in the analysis even if assumptions of a good deal of (probabilistic) 'continuity of development' are a necessary precondition for the analysis.

Another conclusion is that that sort of initial 'demand specification' is extremely important in order to orient the analysis in the labyrinth of modelling work. Otherwise, the more-or-less arbitrary modelling tools may easily control the outcome, and in the end it is the more-or-less arbitrary choice of functions and parameter values which make the output of the model look nice and persuasive. The simplest but at the same time the richest of such demand specifications may be derived from a few stylised facts which the model ought to take into account. In this way the modeller is like a painter who starts by recognising some main features of an object which are to be depicted, and afterwards feels free to ignore many other aspects of it. In the case of Perroux's growth pole theory an important demand specification seems to have come from the nineteenth-century transformation of Continental Europe springing from the heavy industries of the Ruhr district. This case has clear relationships to the main source of stylised facts in this book, namely the process of railroadization.

3.2. Logistic diffusion in post-Schumpeterian terms

3.2.1. Introduction

The logistic equation, proposed by Verhulst and made popular by Pearl, is an attempt to make a simplistic account of typical processes of population growth. It appears a suitable expression for the 'population' of railway tracks. Schumpeter was fully aware of the possibilities of arriving at a 'descriptive trend' (Schumpeter, 1939, 201) through such an attempt to fit the curves. However, he was even more aware of the dangers, especially for aggregate data. He points out that at this level any revealed exponential growth rate has little meaning. He continues:

Still more treacherous and pregnant with danger of speculative temerity may be the application of Verhulst's formula, $y = \dfrac{a}{be^{-t}+1}$, which was intended (1838) to present certain features of organic or of similar types of growth. Even a perfect fit in the least square sense would not prove anything. We are, however, on somewhat safer ground when applying such expressions to the behavior in time of quantities of individual commodities. (Schumpeter, 1939, 492 f.)

But Schumpeter does not apply Verhulst's logistic model or discuss the results.[4] Instead we get a verbal and general account:

[I]t is obvious that ... no industry can go on expanding output at the rate of its innovation stage. Each reaches maturity in the sense that it finds its place in the economic organism and the amount of output beyond which it cannot profitably go, unless that amount be increased by some further innovation within it or in some 'complementary' industry and by the general effects of ... Growth. (Schumpeter, 1939, 497)

Thus we may conclude that:

Output of a new commodity may easily trace out a Verhulst curve which many students will have no hesitation in interpreting as a trend special to that commodity and distinct from any cycles that may run their course in the same period. From our standpoint, of course, this is never strictly correct, although it may, for purposes of partial analysis, be convenient to express oneself so ... (Schumpeter, 1939, 205)

What Schumpeter demands and does not get is the insertion of the development of the 'population' of the individual commodity or sector (like railways) into the overall framework with interconnections with other sectors and with the cyclical pulse of economic evolution. Still, we may ask ourselves whether it was really necessary for Schumpeter to stop here. Later developments have demonstrated that this was not the case. Now we can see that the logistic formula is not only a tool for curve fitting but also one of the starting points for embarking on theoretical analysis, as we see in the case of (evolutionary) ecology (see Kingsland, 1985). Here it is shown that the logistic equation provides us with a rough theoretical scheme which may help to organise the study of several of the issues raised by Schumpeter.

Schumpeter was apparently unaware that Lotka and Volterra in the 1920s had made creative use of the logistic equation as a starting point for their mathematics of interacting populations. Later Goodwin, one of Schumpeter's last students, transferred some of their ideas to the realm of economic analysis. While Lotka and Volterra was studying the interaction between populations of predators and prey (like different species of fish in the Adriatic Sea), Goodwin developed a macroeconomic dynamic model where workers functioned as predators while capitalists had the role of prey (see the papers in Goodwin, 1982). Much more recently 'deterministic chaos' has been found, largely by means of the logistic equation[5] in both ecological and economic subjects. Even the standard exercise for MIT students which had been calculated for decades (The

Beer Distribution Game), was for the first time reported to have surprising properties of deterministic chaos (Mosekilde and Larsen, 1988). In the light of this, Goodwin (1990a) reexamined his earlier studies and created new ones, partly by means of a simple mathematics package for exploring differential and difference equations through computer experiments.[6] These experiences have led Goodwin to suggest a solution to his early attempts to give to Schumpeter formal tools of analysis (see section 1.1.2). Now he emphasises that the complex, 'chaotic' and 'creative' dynamics revealed by a closer study of the logistic equation may help to express Schumpeter's vision of the evolutionary process: 'In my youth and innocence, I tried and failed to persuade Schumpeter to take linear cycle models seriously; in retrospect I think he was right to reject what did not fit his vision, but I think that I could have sold this package to him.' (Goodwin, 1990b, 49) The 'package' of relevance to Schumpeterian thought is of the computer models which include the logistic equation: 'The logistic, which is obviously the most appropriate formulation of [the diffusion of] an innovation, provides the simplest and most direct formulation of the entire gamut of erratic, even chaotic, behaviour of a dynamic system.' (Goodwin, 1990b, 46)

Through his sketches of modelling work, Goodwin (1990a, 14) tries to persuade his readers that new light can be thrown on the economic history of capitalism: its innovations, its growth and its 'spectacular collapses'. Goodwin's idea is to use a (more-or-less modified) logistic equation to represent the investment demand caused by the need to create new production capacity for the exploitation of an innovation. Then the interrelations to the expansion of total demand, output and employment are studied. In this way Goodwin tries to combine Schumpeter's ideas with elements taken from Keynes and Marx (Goodwin, 1986). The work on these ideas is fragmentary but it may be used as an introduction to other kinds of work which tries in a more ambitious way to combine the processes of innovation and diffusion. Furthermore, it may be taken as an example of an explicit relationship between what may be considered the 'demand specifications for new models' as formulated by a representative of the old evolutionary economics (Schumpeter) and the modelling work with a more-or-less explicit relationship to new evolutionary economics.

3.2.2. The logistic difference equation and diffusion of innovations

The yearly increase in kilometres of railway line in different countries may be roughly simulated by means of the logistic difference equation. If we consider N_t to be the number of kilometres of railway line open in a particular country at time t, we may calculate the number of kilometres at the next point of time as

$$N_{t+1} := N_t + rN_t\left(1 - \frac{N_t}{K}\right); \qquad (3.1)$$

where r and K are parameters.[7]

The meaning of these parameters can most easily be explained by explicitly considering the (relative) growth rate of railway line as a function of the total railway line:

$$\frac{N_{t+1} - N_t}{N_t} = r\left(1 - \frac{N_t}{K}\right) = r\left(\frac{K - N_t}{K}\right).[8]$$

The growth rate thus depends on two factors. The first, r, is the maximum growth rate. It can be considered to be a characteristic of the particular innovation (railways) as well as of, e.g., the financial system. The second factor represents the influence of the total number of railway line units on the growth rate. As N_t approaches K, the growth rate becomes smaller and smaller. For $N_t = K$, the growth rate is zero. However, because the expansion occurs in steps, we might see an overshooting of the goal, i.e. $N_t > K$, and a subsequent oscillation around K. For extreme r-values, the pattern of oscillation will become chaotic.[9] The functional dependence of the growth rate on the size of the 'population' of railway units is depicted by figure 3.1.

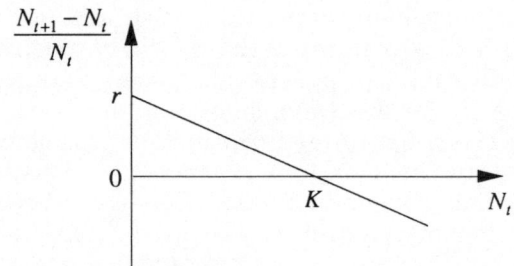

Figure 3.1. The growth rate and the parameters of the logistic difference equation.

The figure may be used to interpret the parameter K as the maximum sustainable level of the application of the railway innovation in a given (national) environment. If this level is not reached, it is still profitable to expand the total length of railway line. If the level is exceeded, parts of the railway network will become unprofitable and more kilometres of line are scrapped than are being constructed. Taken as an abstract model, there are no problems with this interpretation. By doing so, we may simply define the 'carrying capacity' of the national environment with respect to the use of railway-based transport services. However, as soon as we address the historical data as well as make theoretical reflections about the price mechanism, the 'carrying capacity' becomes a concept which is very difficult to deal with. The historical data shows that the railway was part of a complex evolutionary process which led to a sharp increase in the usage of railway services. The price-theoretic controversies in relation to railways demonstrates extreme monopolistic tendencies as well as the heterogeneity of the transport services. We will return to such issues later but it is important to observe that there is no room for them in Goodwin's use of the logistic equation.[10] Here we take as given that the logistic curve adequately describes a density-dependent growth process and thus reflects the aspect of the process of diffusion of an innovation which is most obviously predictable: the environment of each new application of the innovation tends to become more and more densely populated by other applications of the innovation; in the end we have crowding and zero growth. The price mechanism and the set of follow-up innovations may modify this picture but its central message cannot be changed.

Each established railway unit may be seen as a starting point for further railway construction. One of the reasons for this may be phrased in terms of network economies, since established lines may not only compete with new lines but also help them because they increase the total demand for transport services. Another reason why a given railway line may function as a starting point for a new one is related to the fact that awareness of investment possibilities is to a large extent related to direct exposure to the relevant information. Here existing lines may function as a rough indicator of the number of entrepreneurs aware of the possibility of joining the railway business. After a period of 'debugging' which is not taken into account, the effects of the pioneering railway construction on further growth are only dependent on the maximum rate of growth, r. The reason for the diffusion may partly be ascribed to the railway innovation itself, partly to the innovation-promoting agents and mechanisms (entrepreneurs and bankers) and partly to the (expected and real) reaction time of the customers. The complicated time lags in this process are largely ignored. As the established railway kilometres increase, the distance to the carrying capacity, K, becomes smaller. Only the less profitable routes are left, and the increased number of connections in the railway network means that competition becomes more fierce.

Since the tacit knowledge and transfer of information are important parts of the process of diffusion, one will often try to use a model which emphasises spread by contact, as in epidemiological models of medical science. Seeing a process of logistic diffusion as an epidemic emphasises the number of adopters of the innovation, N_t, as well as the number of potential adopters who have not yet adopted the innovation, $K - N_t$. The probability that a non-adopter will become an adopter during the time period from t to $t+1$ is determined by his chance of meeting an adopter as well as of the 'potency of spread' of the innovation. These expressions have a very simple interpretation in many diffusion studies of, e.g., fashion but they may not fit the railway case.

3.2.3. Short and long innovation-based business cycles

In the 'weird and wonderful world of chaos' (Goodwin, 1990b, 46) there is little room for believing that we can find empirical business cycles with any strict periodicity. However, there is plenty of room for Schumpeter's idea that capitalist evolution takes place within a loosely defined cycle. Such cycles may be the outcome of a logistic diffusion process of an innovation or, rather, a 'swarm' of innovations. In the following we will see how this idea can be developed both generally and with some relation to the stylised facts of the Schumpeterian standard case of railroadization.

The argument about cycles is based on assumptions about the innovation under study as well as the economic system into which it is introduced. These assumptions are not systematically stated by Goodwin (1990b; 1990a) but they may probably be summarised as follows, as also seen in the characteristics of the epoch of railroadization. First, implementation of the innovation presupposes relatively huge investments in physical capital because of a small output-capital ratio. Second, there is a potentially huge demand for the services provided by the innovation. Third, the core innovation is related to a 'swarm' of other innovations which in general have small output-capital ratios. Fourth, entrepreneurial and financial attention is concentrated on this swarm of innovations at the expense of a huge number of other and potentially profitable innovations for a long time. Fifth, the investment capacity of the economic system is relatively limited compared with the investments necessary for the implementation of the swarm of innovations. Its implementation is thus a question of decades. Sixth, each swarm of innovations consists of some innovations which can be almost fully exploited during a 7 to 10 year diffusion process and some which will only reach their long-term K-level through, e.g., a 50-year period of diffusion.[11]

These assumptions may be combined with 'vaguely plausible parameters' (Goodwin, 1990a, 19) to make simulation models of this kind of an evolutionary process. We may start with an 8-year diffusion process for a smaller 'swarm' of innovations. In other words, it takes roughly 8 years for a process of investment in innovations to reach the carrying capacity of the economic system (with roughly given input and demand conditions). In relation to equation 3.1 it means that we have a K-level which for plausible values of the rate of growth of capital embodying an innovation, r, can be reached within that period. The result may look like figure 3.2.

Density-dependent diffusion and innovation 73

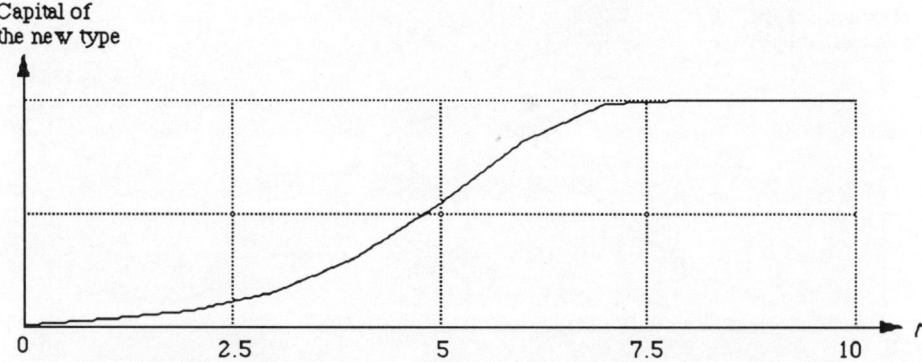

Figure 3.2. 8-year expansion of the capital embodying an innovation.

We now turn to the effect of investment embodying the new innovation on total demand and output. For a rough discussion of this issue we assume that investment in the (group of) new innovation relates to an economy with a given process of change. It may be stationary, growing, moving randomly, or in cycles but we assume that its movements do not influence the diffusion process. To start with, interest is concentrated on the added effects on output, expenditure, and income. To see these effects, we superimpose equation 3.1 on the economy in the most primitive way. The new-type capacity, N_t, starts at zero at time $t = 0$ and comes very near to K at $t = 8$. The corresponding capital is found by multiplying the capacity by κ, the capital-output ratio. The increase in capital, $\kappa(N_{t+1} - N_t)$, represents the extra effect of the innovation on demand in period t. The extra demand created in earlier periods has also an effect through its effect on output. The process which adjusts demand and output is assumed to be quite simple. The output is adjusted by a certain proportion, α, to any discrepancy between output and demand. Given the propensity to spend a proportion, δ, out of the extra receipts of Q_t we have something like

$$Q_{t+1} := Q_t + \alpha\big((1-\delta)Q_t + \kappa(N_{t+1} - N_t)\big); \tag{3.2}$$

which may be studied with 'plausible' but highly simplifying parameter values. Thus we shall simply set $\alpha = 1$ which means a complete adaptation of output to demand during one period. Using different scales for the new-type capital and the extra output we arrive at figure 3.3. Here we see the cyclic effects on output of the diffusion of a small swarm of capital-demanding innovations.

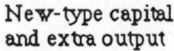

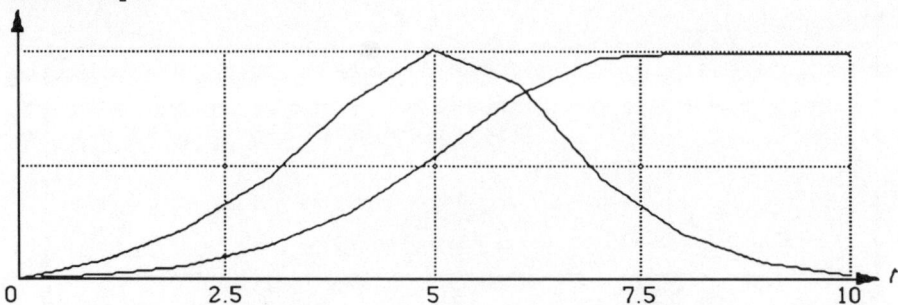

Figure 3.3. Effects on output of the capital-demanding diffusion of an innovation (different scales).

It is clearly possible that there is a relationship between the logistic equation and the cyclical behaviour of the economy. However, there are several problems with the Schumpeter-like evolutionary step depicted by figure 3.3. First, we have to explain why the swarm of innovations is really a swarm and not a sequence of randomly occurring logistic curves with no overall effect on aggregate output. At present we simply assume this behaviour. Second, we have to assume that innovations have a high capital-output ratio or have large-scale indirect effects on investment behaviour. The railway is a plausible example of such an innovation, but it is difficult to generalise from this analysis (as emphasised by Schumpeter). Third, we observe that the level of output is back where we started after a 10-year period while the productive capacity is increased with a potential of supplying more, e.g. transport services. This is clearly an Achilles heel of the Schumpeterian set-up which becomes even clearer when we study the relationship between output and employment under the assumption of increasing labour productivity. In Goodwin's (1990a, 21) opinion the only way out is some Keynesian-like mechanism. However, an emphasis on demand enhancing innovations may be a more Schumpeter-like supplement.

The history of capitalist evolution may be depicted by a series of such innovation-based cycles. However, according to Schumpeter there is more structure to the overall history of capitalism. Parts of this structure are reflected by the stylised history of railroadization (see chapter 2 and Andersen, 1993a). These are the long waves of capitalist evolution which Schumpeter (Schumpeter, 1939) named Kondratiev waves. In the beginning of such waves the shorter (Juglar) cycles have more and more boom-like upswings while the downswings are weakening. In the later parts of the long waves the situation is reversed. Taking the above results for given (see equations 3.1-3.2 and figures 3.2-3.3) the problem is introducing a long cycle. Goodwin's simplified solution is to combine a series of shorter diffusion processes with a long-term diffusion process.

Such a solution raises several difficult problems. For example, the interdependence between the shorter and the longer term diffusion processes poses complex problems. To illustrate some of these problems, we examine a well-established 8-year cycle and focus upon the effects caused by adding a logistic diffusion process of about 50 years. Each cycle has its own more-or-less well-formed structure.[12]

Assume that we still have a basic economic process and that we are only studying the changes created by the shorter and longer diffusion processes. Furthermore, we assume that the shorter diffusion process creates a cycle which is already understood and that it may be modelled as an 8-year cycle with constant amplitude. This may simply be done by making the (added) output Q_{t+2} depend positively on the output in the previous period and negatively on the output two periods ago. In the computer simulation we may add a constant component of effects of investment, Δ^{short}, in order to make the short cycle robust with regard to the subsequent manipulations.[13] This cycle is then interrelated to a long-term process of logistic diffusion. On the one hand, the output of $t+2$ is related to the investment demand created by the innovation-based sector in the previous period (equation 3.3).[14] On the other hand, we have a simple long-term diffusion process which is basically the simple logistic difference equation (3.1). However, the capacity of the capital implementing the long-term innovation, N_{t+2}^{long}, is coupled to the output of period $t+1$.

The two-cycle process is given by the following equations:

$$Q_{t+2} := aQ_{t+1} - bQ_t + \Delta^{short} + \kappa\left(N_{t+2}^{long} - N_{t+1}^{long}\right);[15] \qquad (3.3)$$

$$N_{t+2}^{long} := N_{t+1}^{long} + r^{long} N_{t+1}^{long}\left(1 - \frac{N_{t+1}^{long}}{K^{long}}\right) + cQ_{t+1};[16] \qquad (3.4)$$

The result of the computations defined by these equations (or assignment statements, see the following chapters) is, of course, heavily dependent on the choice of parameter values. They allow an extreme variety of patterns, but they can generate the pattern depicted by figure 3.4 by the following crude method: First, we set parameters a and b in a way which creates a short cycle of about 8 years and therefore about 6 cycles per 50 years. With vaguely plausible values for κ and c, we then search for a logistic process of expansion of capital which implements the slowly diffusing innovation (or innovative swarm). K^{long} must have a value which influences the shorter cycle, and finally we set r^{long} to a value which gives a 50-year process of diffusion.[17] After some experiments with the parameters we find figure 3.4.

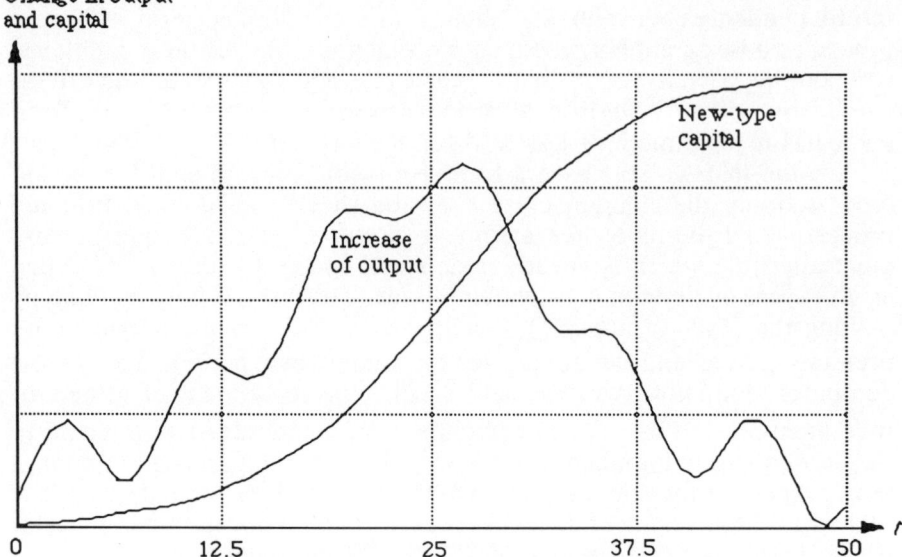

Figure 3.4. A 50-year diffusion of an innovation, interfering with six 8-year innovative cycles (different scales).

The long process of diffusion is clearly able to influence the regular 'Juglar' cycles in a way which illustrates some aspects of Schumpeter's idea of the long cycle. In the beginning the long-term expansion of the capital embodying an innovation (like the railway) is more and more emphasising the upswings and weakening the downswings. Therefore, the short cycles become more and more characterised by expansion and less and less by contraction. Later this effect disappears. After the fourth peak we see a drastic contraction which is interrupted by a 'hump' which can only barely be taken for a fifth peak. The period ends with a 'normal' shorter-term cycle.

3.2.4. Some limitations of the simple logistic-curve approach

The above presentation only covers a small fraction of the issues which Goodwin wants to develop. His concern with productivity and employment can easily be added but not the chaos-theoretic context in which the argument is developed. Goodwin's use of the logistic equation must be evaluated with respect to Schumpeterian thoughts. In particular, we remark that Goodwin assumes that the process of innovation is totally exogenous. In his equations, economically relevant knowledge is assumed to grow steadily (by e.g. 4% per year). Exploitation of this knowledge is, however, taken to come in waves which can be explained in several ways.[18] However, because the process of innovation is exogenous,

Goodwin does not develop a discussion about the relative weights of further innovation and logistic diffusion.

As one way to develop these ideas, Iwai's (1984/93; 1984) account is quite interesting, although it is stated in terms of an industry rather than an economy. Like Goodwin, Iwai is fascinated by the logistic curve but he follows Nelson and Winter in insisting on the technological heterogeneity of the industry (economy). This means that he can apply 'the logistic law to the description of the evolutionary pattern of the whole array of production methods co-existing side by side at the same time.' And this strategy allows him 'to study the dynamic interaction between processes of imitation and innovation in an integrated manner'. (Iwai, 1984/93, 134) In Iwai's drama the economy may reach a situation where diffusion of the available innovations has been completed, and all firms are applying the best-practice technique. This situation, however, creates strong motivations for innovation:

The tendency towards technological uniformity among firms is bound to be upset by a sudden introduction of a new and better production method by one of the firms. Indeed, to destroy the stalemate brought about by the imitation process and to create a new industrial structure is the role our capitalist economy has assigned to Schumpeterian entrepreneurs or to innovative firms. (Iwai, 1984/93, 135)

What is lacking in Goodwin's work but is present in Iwai, Nelson and Winter, and other evolutionary researchers is what we call 'population thinking' (see section 1.1.3). This expression from evolutionary biology builds on the heterogeneity principle, and we have already seen how it is possible to operate with a 'population' of transport firms which have different ways of providing similar transport services; Schumpeter's favourite illustration was long-distance transport services based on mail coaches and railways. However, even these two groups are still heterogeneous in important respects which have to be taken into account in some analyses (see below). To develop mathematical models, we make general distinctions between larger classes or populations of comparable routines, j, each of which consists of one or more routine types, i.

The introduction of population thinking is one of the ways of overcoming the fixed parameters of the logistic model. The fixed parameters may be considered as characteristics of subpopulations and the changing frequencies of these subpopulations change the parameter values of the aggregate population. Nevertheless, even this formulation does not fully fit the Schumpeterian vision of capitalist evolution. At least implicitly he suggests that a hypothesis about fixed parameters (or any fixed functional form of the change of r and K) may serve to hide interesting aspects of the cyclic-evolutionary process of, e.g., railroadization. The modern answer would probably be that although this is correct, the logistic interpretation still provides an analytical framework of considerable heuristic value if we accept that there are many possibilities for the parameter to change during the process. The parameter r_t may change in a very complex way from the introduction of the innovation, t, to the point where we consider it to be 'fully spread', $t+n$, because, e.g.:

1. The railway innovation is undergoing change.
2. The types of entrepreneurs, managers and bankers involved in the railway business are changing.
3. The macroeconomic conditions are changing.

Similarly, K_t may change because of, e.g.:

4. Quality improvements and cost reductions in relation to the innovation.
5. Developments in competing and cooperating sectors.
6. Changes in the macroeconomic variables.

In this perspective we may think that r_t and K_t increase over time in the first part of the process. Such increases may be due to the fact that the first railway entrepreneurs and bankers pave the way for their followers (network externalities, specialised suppliers, new methods of finance, etc.). Furthermore, the railway pioneers themselves help to move the frontier, the point where saturation is reached. Without this effort there might be much too little demand. Thus, the pioneers did not adapt to existing K_t's; instead they recreated a 'new world of economic norms' and acquired demand for their 'innovation'. This is especially clear in new regions where, e.g., the US created the demand for its own services. (Schumpeter, 1939, 319)

All in all, it seems that in Schumpeter's mode of thinking the apparent parameters of the logistic model are central variables of another model which is not formulated sharply and which appears to have elements which are 'refractory to mathematical formulations' (Schumpeter, 1991a, 230). The reason is basically that we are essentially trying to describe an irrevocable process which is dependent upon the timing and character of events to such a degree that the model should, actually, work in 'historical time'. Taken literally this is, of course, an impossibility. As a more metaphorical statement it may help to encourage modellers to be more sensitive to the stylised facts of history.

3.3. Lotka-Volterra inspired models of interdependent diffusion?

3.3.1. Fables of density-dependent evolution

New routines in the provision of transport services interact not only with old routines in transportation but also with routines of many other areas of economic life. To study this and other cases we need a classification of types of interaction or 'coevolution' of different routines, and this taxonomy is most easily developed by studying pairs of different routines (see May, 1976/81, ch. 5; Roughgarden, 1979, ch. 21). Here, we shall compare two types of situations: one where the commodity based on the routine is *alone* in a segment of a given (market) environment, and another where different commodities (considered as carriers of routine types) coexist.

Before we proceed to our series of approximations to Schumpeter's thought, a couple of fables can introduce a more intuitive understanding of the following discussion. To begin with, we can relate to a major source of modelling tools and ideas, namely a kind of Robinson tale of evolutionary biology.[19] This illustration is not wholly arbitrary since the study of island biogeography has played an extremely important role in evolutionary analysis from Darwin's visit to Galapagos and Wallace's studies in Indonesia to present-day developments of central ideas (MacArthur and Wilson, 1967). Let us take an island which has plenty of vegetable life but no animal life. Suppose that the first animals which arrive are a few rabbits. They bring with them a set of genes with particular frequencies of occurrence. Provided the rabbits come from similar environments, the frequencies will not change (much) during the next generations and thus there is, per definition, (practically) no evolution. Nevertheless, the island is filled with life and with change which may be described crudely by means of the logistic difference equation (3.1). Thus, the rabbits will multiply rapidly (high r) but increasing mortality will put a brake on growth until their numbers reach the island's carrying capacity (K) and net growth has come to a halt, or alternatively, moves in violent cycles around this level.

There are different ways of introducing evolutionary disturbance into this system. In principle, a major and dominant mutation could take place among the rabbits and this would create a period of change in gene frequencies. However, a more likely disturbance would involve a few pioneering wolves who arrive at the island. They bring with them their own genes. Furthermore, the newly arrived wolves will probably start a radical change in the gene frequencies of the rabbits, especially if the rabbit population has not met predators for many generations. Under this condition evolution is created by an inter- and intra-species interaction (predation and competition). Among the rabbits, survivors will, e.g., hear relatively well, dig deeper, etc. Among the wolves there are years of rabbit shortage and special genes will be present in many of the survivors. After some predator-prey 'cycles" (with changing parameter values) the movement of the distribution of gene frequencies will come to a halt (provided that one or both species do not die out and important mutations do not occur). Even so, cycles will probably continue and a crude description of them may be given by the classical predator-prey model (with fixed parameters) described by the Lotka-Volterra difference or differential equations (May, 1976/81, ch. 5). On the other hand, the fixed parameters of the predator-prey model exclude (most types of) evolutionary change in the frequencies of genes in the population

This story of animal Robinsons and their progeny at the desert island corresponds fairly well to the Schumpeterian scheme (even if Schumpeter disliked biological analogies, see Hodgson, 1993, 146). The wolves represent the creative-destructive force of the innovators. However, one aspect of the Schumpeterian vision is clearly lacking: the self-generated change of the economic system from a non-evolutionary to an evolutionary state. It is next to impossible for a new species to invade a mature ecosystem, and much less to disturb the equilibrium position of the gene frequencies seriously. The reason for the success of the rabbits and the wolves is precisely because the island has a very immature ecosystem. In a well-established and mature ecosystem the only way to introduce a shock is through exogenous forces. This is precisely what Schumpeter wanted to avoid.

After this simple example we are prepared for a fable about the process of railroadization. In order to avoid studying the period involving the initial emergence and debugging of the innovation, we make an artistic shortcut: In a first step we develop an equilibrium situation by colonising country A(merica) with pioneers who are equipped with well-suited routines, including norms for producing horse-driven mail coaches, principles for organising transport systems and a lot of other economic routines which are adapted to the possibilities and relative costs related to mail-coach-based transportation. Alternatively, we may give the selective forces time to adapt the character and frequencies of the routines to the conditions of the new land. We assume further that there is a period where the economy of A is functioning in a way which does not challenge the routine frequencies. Even cycles have been recognised as phenomena which do not indicate a need for changes in routines.

As a second step we assume that some economic agents who have not forgot the pioneering spirits introduce railway projects (which they have learned of in country B(ritain)). They know the routines and the related price system (or the limits within which prices may change). They foresee a railway-based system of future costs and prices[20] which is clearly to their advantage. They are even entrepreneurs in the sense that they are able to persuade others to finance their projects and later to buy the services of the newly constructed railways. Their subsequent successes lead to further railway projects and in the end there is a whole 'swarm' of entrepreneurs promoting railway construction (and other changes). Together they force the other agents of the economic system to begin to realise that the situation has changed and that things are in a flux where the relative frequencies of routines are changing and will continue to change.

The third step is rather confusing. Some actors do not react at all. Others, like the owners of mail coaches, are forced to do so. Still others make adaptive changes in behaviour in order to exploit the possibilities created by the disequilibrated situation. Many of the activities combine to recreate a tendency towards unification and equilibrium. Even so, in the prosperity created by the railroadization there is no strict necessity to react to the new times. However, when the recession sets in, maladapted firms must liquidate and rapid changes in the frequencies of routines take place. The character of the railway innovation makes it impossible to find the equilibrium in a single 'cyclical-evolutionary step'. Several cycles are necessary before the economy of country A settles down to a new state where evolution has come to a halt. But does it really settle down? To Schumpeter, Iwai, and many others the answer is no.

3.3.2. The interdependence of applications of different innovations

The simple account in the above fables in terms of a density-dependent diffusion process is not sufficient. It only holds where an innovation is applied in a new environment where it has no (significant) competitors or other types of interactions. Here the growth rate is expected to change in a density-dependent way and to approach zero as the number of applications of the commodity approach the carrying capacity of the particular environment. Apart from the lonely rabbit population, we have not met such a case in the fables. Instead, we have met interdependence, especially competition. The task is now to discuss such cases by means of a very simple case where two 'populations' (like mail coaches and railways) are influencing each other's carrying capacities. This interaction may be described in terms of the Lotka-Volterra equations of coevolution; these equations may be considered extensions of the logistic equation in order to cover cases where the applications of different types of routines are influencing each other. As usual we shall limit our discussion to difference equations even if most of the studies of the Lotka-Volterra equations are made in their continuous form which is easier to analyse mathematically. At the moment we concentrate on the simplest possible expression of the interaction of the relative growth rates of routine-1 and routine-2:

$$\left.\begin{array}{l} N_{1,t+1} := N_{1t} + r_1 N_{1t}\left(\dfrac{K_1 + a_{11}N_{1t} + a_{12}N_{2t}}{K_1}\right); \\[2ex] N_{2,t+1} := N_{2t} + r_2 N_{2t}\left(\dfrac{K_2 + a_{22}N_{2t} + a_{21}N_{1t}}{K_2}\right); \end{array}\right\}^{21} \quad (3.5)$$

where N_{1t} and N_{2t} are the number of applications of two routines (e.g., numbers of commodity units); r_1 and r_2 are the intrinsic growth rates (potencies of spread) of the two routines; K_1 and K_2 are the carrying capacities of the two routine applications if left alone; a_{12} is the interaction coefficient for the effect of an individual application of routine-2 on the expansion of applications of routine-1, and a_{21} indicates the opposite direction of influence; the signs of the a's have been chosen to facilitate the following discussion.

This system of modified logistic equations is closely related to the one-routine case (equation 3.1). In this case the relative growth rate of the applications of routine-1 becomes zero if the carrying capacity, K_1, is reached. In the case of two interacting routines, zero growth is obtained at another level of the routine applications. In the case of inter-routine competition, it is reached while N_1 is still less than K_1; in the case of symbiosis, growth continues even if the population has expanded beyond K_1.

When trying to interpret these equations of bilateral interaction, the Lotka-Volterra equations,[22] it is useful to remark that the density-dependence of each of the populations can be considered as a result of intra-routine competition. The strength of this intra-routine competition is normalised to 1 ($a_{11} = 1$ and $a_{22} = 1$). In this context $-a_{12}$ and $-a_{21}$ can be seen as indicators of the strength of inter-routine *versus* intra-routine interaction, or, in other words, the strength of interaction between different routine types *relative* to the interaction within a given routine type. Here we have different cases which may be classified as competitive, mutualistic, indifferent, etc. Among the competitive interactions we have the special case where a_{12} is equal to -1, i.e., inter-routine and intra-routine competition are equally strong. In this case the dampening effect of an extra application of routine-2 on the expansion of the applications of routine-1 is equal to the effect coming from an extra routine-1 application. Another type of case is found when $-1 > a_{12} > 0$, i.e., the inter-routine competition has less effect than intra-routine competition. But there is also the possibility that $a_{12} < -1$, i.e., the effects of inter-routine competition are greater than the effects of intra-routine competition. If $a_{12} = 0$, there is no interaction. If $a_{12} > 0$, we have not competition but mutualism since the effect of routine-2 is to expand the carrying capacity of routine-1. These and other effects are indicated in table 3.1.

Table 3.1. Types of bilateral interaction.[23]

		a_{21}		
		+	0	−
	+	+,+	+,0	+,−
a_{12}	0	0,+	0,0	0,−
	−	−,+	−,0	−,−

Interaction of the two routine types can now be discussed in terms of a_{12} and a_{21}. The case where both are zero is trivial. This leaves us with three possible combinations of signs for a_{12} and a_{21}. If both are less than zero (the '−,− case'), then we have what is normally considered to be competition. If both are greater than zero (the '+,+ case') we have mutualism (which may be extended with the '0,+ case' and the '+,0 case'). If $a_{12} < 0$ and $a_{21} > 0$ (the '−,+ case'), or if $a_{12} > 0$ and $a_{21} < 0$ (the '+,− case'), then we have asymmetric interaction of the type which in biology is called predator-prey interaction. However, because of the idiosyncrasies of the logistic and the Lotka-Volterra equations it should be remarked that we have here defined special versions of the broader concepts of competition, mutualism and predation.

The generalised Lotka-Volterra equations and the related classification of interactions will turn out to be very important when we try to describe and explain movements in relative routine frequencies or in absolute routine frequencies. To give an immediate impression of the issues at stake, I will shortly discuss some of the two-routine cases in terms of movements and equilibria in their phase-plane.[24] Each point in the diagrams of figure 3.5 (see next page) represents the relative size of the application of routine-1 (N_1) and routine-2 (N_2). Starting from a specific point, we can see that in the next moment each of the routines may increase, decrease, or be unchanged. The case of increase is marked by hatching and the case of zero growth by lines. In the non-hatched area the routine applications are decreasing. The movements in the phase-plane end up in stable equilibrium points where both rates of change are zero and where small disturbances cause a movement back to the equilibrium point. These points are marked with dots. By treating the '−,− case' of competition in these terms, we see that it falls into four subcases (figure 3.5).

There is a basic difference between cases (a)–(c) and case (d). In the first three cases we see that one of the two routines will become totally dominating. In case (a) it is routine-1 which wins and in case (b) it is routine-2. In case (c) the outcome depends on the initial conditions. Given certain conditions routine-1 will win, but under other conditions routine-2 will win. These cases are straightforward and represent the idea that competition must result in one type of winner. If the innovation has sufficient fitness it will win over its competitors as railways won over mail coaches. However, there might be cases of path-dependency where the innovation will only win over its competitor if it reaches a certain proportion relative to the competitor. Take, for example, the two competing standard measures of the distance between the rails, i.e., gauges. The Stephensons chose a five feet[25] solution based on mining traditions (dating back to the Romans) while others argued that a seven feet gauge was faster and safer. The latter solution was crowded out by the former (Simmons, 1978, 26–28, 45–47, 84 f.).

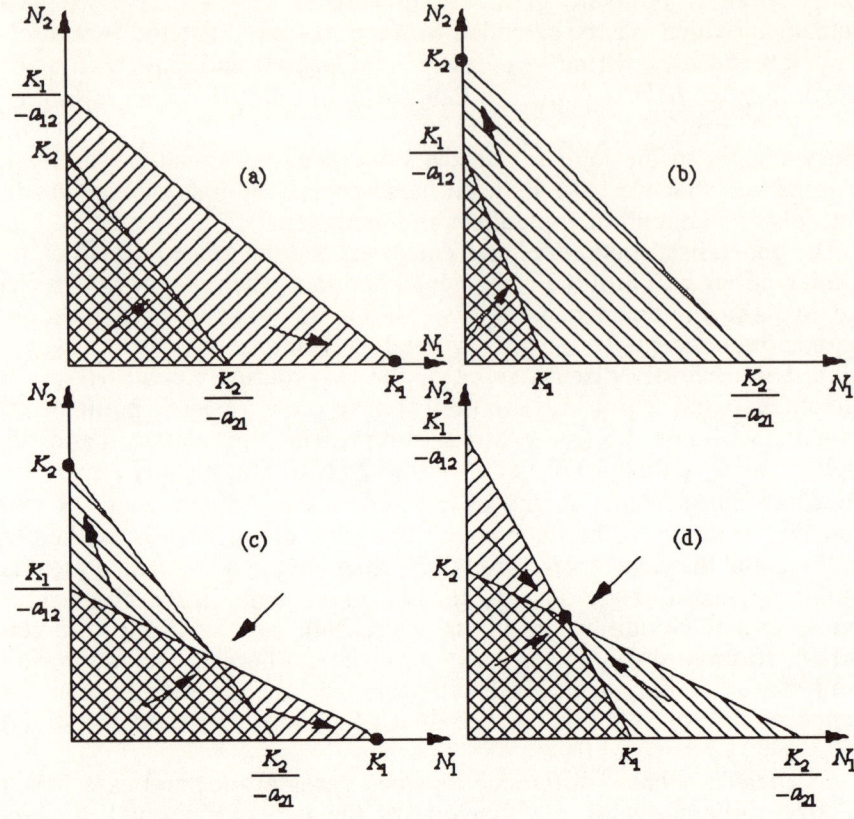

Figure 3.5. Competitive interactions between two routine types.

The line $K_1, K_1/a_{12}$ represents the number of applications of routine-1 where its rate of change is zero at various densities of routine-2. The line $K_2, K_2/a_{21}$ represent a similar number for routine-2 at various densities of routine-1. The dots, •, represent stable combinations of N_1 and N_2.

▨ N_1 increases, ▧ N_2 increases, ▩ N_1 and N_2 increase.

3.3.3. Evolutionary interaction and niches

On the basis of these cases, we may formulate the principle of competitive exclusion which simply states that 'complete competitors cannot coexist' (Hardin, 1960/75, 35). This principle is constructed in terms of four ambiguous words to emphasise that it contains a heuristic for further research rather than a 'law'. In the present context an immediate task seems to be to deal with case (d) which appears to contradict the competitive exclusion principle. This case is constructed by means of equations 3.5 and such values of a_{12}, a_{21}, K_1 and K_2 that routine-1 and routine-2 are able to coexist while at the same time conforming to our immediate conception of competition. The condition for coexistence is that when each type of routine is very abundant, it inhibits its own further increase more than it inhibits the other's. This is the case in case (d) where routine-2 increases in the area where routine-1 has transcended its individual carrying capacity and is decreasing. Similarly, we see that when routine-2 goes beyond its carrying capacity, we come to a zone where routine-1 increases. In both zones the increase of either routine inhibits its own growth more than it inhibits its competitor. Thus, the currently weaker routine gets a chance of regaining its strength. The result is that neither routine is able to drive the other out of economic life. The explanation may be that further application of the two routines is limited by different selection environments. In light of the heterogeneity principle with respect to both the individual applications of a routine type as well as the selection environment, this possibility is obvious. Within the two routine types there may originally be individuals with almost identical characteristics and here the fittest routine crowds out the weaker one. Even so, other subroutines within the two routine types differ more sharply and they may coexist. In other words, the selection pressure created by the competition forces the routine types to become more specialised which reduces inter-routine competition. The limiting factor is instead the narrowness of the market niche into which each of the competing routine types is forced.

In relation to the 'spread' of an innovation like the railway we find plenty of examples of specialising effects of competition. Take the relationship between railways and canals. In the beginning, the two were seen to be directly competing and the monopolistic rents of canal owners were undermined. However, even late in the nineteenth century when huge productivity gains had been obtained in the railway system, canals had in many parts of the world freight rates that were about half of the rates of railway freight. There was a difference. Canal transport was very much slower and in many areas ice prevented transportation for several month each year. So, there was some room for both types of transportation. Even more drastically we see the same in the case of railway–mail coach competition; but applying the principle of competitive exclusion while ignoring the possibility of specialising competition and symbiosis led to early predictions for the UK that were radically wrong. While they concluded 'that the use of 1,000,000 horses would be superseded allowing for the subsistence of 8,000,000 human beings, the railway instead had the effect of increasing the demand on the horse'. (Berg, 1980, 30) The reason was that horse traction was used for short-distance transport, partly to the railway. For example, the railway was central for Sherlock Holmes' extension of his detective business to broad areas of Britain, but he had to take a horse-driven hansom cab to Victoria Station.

However, coexistence is not a sign of lack of competition and we cannot conclude that Schumpeter was wrong in emphasising the jump from mail coaches to railways. The destruction of long-distance stage coach know-how and related firms is a very real thing. In some cases it looks less like competition and more like predation (the '–,+ case') since the expansion of stage-coach activities in a certain period may have pointed out new markets for railway services by indicating profitable routes. Their expansion probably also helped railway expansion, while the subsequent railway expansion destroyed the stage-coach lines. Similarly, we see that the fabulous Pony Express from Missouri to California helped to catalyse the construction of the first transcontinental telegraph line which in turn stopped horse-based mail service. At the same time, the railway developed in symbiosis with other types of horse transport. The relationship of mutualism is even clearer between the railway business and industries like steel and locomotives; the expansion and even the productivity of this cluster of industries are dependent on its different members. As long as this 'development block' (Dahmén, 1950/70) was able to create expansion in, e.g., the demand for transport services it also created broader 'niches'. At a lower level of demand such niches would have been too small to carry a specific routine type but as railroadization spread, they became sufficient for an increase in the diversity of transport services and industrial production.[26]

This example of an evolutionary change in the parameters of interaction of the Lotka-Volterra model covers only a fraction of a huge topic. Despite that, there is a basic lesson to be learned: we can find a heterogeneity of potential evolutionary importance at practically any level of aggregation we can imagine. During an epoch of economic evolution this heterogeneity below our accustomed level of discourse may be of little relevance. When the conditions and the selection environment are radically changing, we need to return to a lower level of aggregation and see whether we have to reconstruct our aggregates. Here the logistic model and its extension into Lotka-Volterra-like models may be of much help in focusing on certain questions.

However, much caution is needed. The logistic curve is stylised economic history rather than a tool for picking the winners. It reconstructs this history by means of two sets of parameters (the rs and the Ks) which cannot always be given a precise economic meaning. Furthermore, it is influenced by the conditions of biological reproduction where the rs are relatively autonomous attributes of species and their variants while interaction is related to the Ks. There are other formulations; through the fable of rabbits and wolves we know that the reproduction rate of wolves is correlated to the abundance of prey. In economic affairs this kind of interaction is extremely complex and brings us far beyond the reach of the primitive models discussed up until now. However, such models help to focus our attention on a problem: interaction which does not influence the Ks significantly but which has an important effect on the rs. This type of interaction is probably wide-spread in our case of railroadization. The underlying mechanisms are closely related to economic decision-making and the financial system.

3.4. The logistic curve and economic decision-making

3.4.1. Firm-level analysis

Ecologically inspired analysis is normally characterised by its relatively high level of aggregation. Because of this, individual behaviour is not always adequately reflected in the parameters but this is one of the costs of the increased power of expression implied by the general formulations. This hinders the development of evolutionary-economic analysis. Here we are basically dealing with economic decision-making which is explicitly concerned with the future effects of present-day action. In this setting we may introduce the perceived distance from the saturation level, K, as a decision-making variable. This section suggests a few possible ways of developing this idea. However, we shall first (in relation to Lloyd et al., 1975) see how expansion of the output of a couple of firms can be described in terms of density-dependent growth.

Let us start with a firm, A, which knows something about the basic parameters of the market and its own possibilities. There is a minimum scale for making profitable production, M_A, and there is a maximum or saturation level for the firm's growth, K_A. The first parameter is new in respect to the previous discussion. This parameter is clearly seen from the firm's view about a relatively established innovation (or a small and predictable innovation). If the firm is unable to reach this level, it will sooner or later go out of business. If it transcends the level, it will for some time have increasing returns. Let us first take the case of a monopolist which may expect to have possibilities of profitable growth between M_A and K_A:

$$N_{A,t+1} := N_{At} + r_A (N_A - M_A) \left(\frac{K_A - N_{At}}{K_A} \right); \qquad (3.6)$$

where N_A is the output of the firm (interpreted as representing applications of routines); r_A is the maximum growth rate that the firm is able to obtain, given the conditions of investment and finance; K_A the carrying capacity and M_A the minimum efficient scale.

Equation 3.6 needs to be interpreted in terms of profitability. Profits may enter into the argument in complicated ways (as Schumpeter was all too aware). One of the simplest possible ways is in line with the kind of adaptive response we have been focusing on in the present chapter:

$$N_{At} = \beta_A \Pi_A (N_A),$$

where β_A is a positive constant characteristic for the firm and $\Pi_A(\ )$ is the excess profit function of the firm. This equation simply says that the firm increases its output of products which have a profitability above a certain level. If this level is not reached, the output is contracted. With this simple behaviour, the main problem for the firm is to find how far or how close it is to K_A.

As soon as we take potential entry into account, the problems get more complicated. To the firm's strategists an increase of the minimum efficient scale $(M_A + m_A)$ may enable them to influence the possibility that would-be competitors will enter. First, it would signal to them that the monopolist has made a credible commitment to defend a higher proportion of its market share. Second, it may influence the product in a way which also increases the minimum efficient scale of new firms in the industry. We shall not include this possibility explicitly but simply turn to the general form of (competitive) interaction between two firms, both in reality and when a firm, B, is considering entry to the routine/industry. Here we may use the Lotka-Volterra model (equations 3.5), but first we should note that the products produced by two firms in an oligopolistic situation are never exactly the same (because of trade marks, patents, etc.). In this case we have a partially competitive situation which may be described by adding the effect of the one firm's output on the expansion of the other firm's output, and *vice versa*. To make the parallel to equations 3.5 obvious we also make the interaction coefficient of the firm's own input explicit ($a_{AA} = a_{BB} = -1$). Thus we have the following assignments:

$$\left.\begin{aligned}N_{A,t+1} &:= N_{At} + r_A(N_A - M_A)\left(\frac{K_A + (-1)N_{At} + a_{AB}N_{Bt}}{K_A}\right); \\ N_{B,t+1} &:= N_{Bt} + r_B(N_B - M_B)\left(\frac{K_B + (-1)N_{Bt} + a_{BA}N_{At}}{K_B}\right);\end{aligned}\right\} \quad (3.7)$$

where subscripts A and B refer to the two firms' activities with respect to a given but not homogeneous product and N is the output of this product; K is the carrying capacity should one firm be left alone; a_{AB} is the effect of a product unit produced by B on the change of A's output.[27]

3.4.2. Pioneering and crowding modes of behaviour

An increase in the heuristic value is created by using the logistic curve as a framework in which to interpret the economic decision-making related to the process of railroadization. According to this interpretation we are dealing with the expected values of the presumably logistic expansion at different future points of time and the effects of these expectations on, e.g., new construction of railways and complementary investments. This interpretation is suggested by Schumpeter's formulations about railways being 'overdone', that is, railway companies securing demand by promoting other investments.

Let us take the viewpoint of a firm at a particular point of time, t. The firm has expectations about the growth of output until a point of time, $t+h$, which is determined by its 'economic horizon'. First, we assume that the firm is basing its expectations on the hypothesis that output is expanding in a logistic manner. Second, we assume that the firm has hypotheses about the approximate values of r and K for the whole industry. Even so, there is little reason to believe that the resulting expectations are formed in any precise manner. The relevant investment criterion for firm A is instead expressed in terms of whether the railway business at point $t+h$ will be 'far away from' or 'close to' its theory about the density-dependent modelling, carrying capacity of the industry K^*. This carrying capacity is probably defined in a way which puts emphasis on freight rates and quantities sufficient to uphold a minimum level of profitability. Whether this level will be reached at point $t+h$ depends on the hypothetical r^*. Making these judgements involves expectations about other sectors and commodities and even about different macroeconomic considerations.

It is such manifold problems of judgement which distinguish the Schumpeterian entrepreneur[28] from the Schumpeterian manager. The manager prefers a well-established situation where the r's and K's are well-known and this situation can mainly be found if the system is close to its K-levels. The innovative entrepreneur is characterised by Schumpeter as operating in a totally different situation, far from the K-equilibrium. We may thus classify different strategies of economic behaviour in relation to the selection processes in uncrowded situations (with room for r-determined expansion) and crowded situations (near the K-limit). In the early stages of the railway diffusion, selection favours fast action and spread, i.e., pairs of entrepreneurs and financiers able to provide a high r-value. In this situation we see the relevance of Schumpeter's dictum that 'one must be right at given dates.' (Schumpeter, 1939, 412) One of the problems of timing is related to the fact that fast expansion is normally related to financial vulnerability. If the recession of the economy comes too early, failure of the project may follow. Still, the prospects of rapid creation of a strong position keep up the 'animal spirits'.

Later, when the railway system is well-developed, speed is dangerous and does not lead to any results. What is needed is the fine-tuning which may give a lead *vis-à-vis* the competitors or which may marginally move the K-frontier. We have seen examples of later developments in the above discussion. This type of behaviour is especially important during long-term downswings where the sustainable level may have fallen below the existing industrial capacity:

Rationalization ... expresses the gist of what we mean by downgrade developments: exploitation to the utmost, partly under duress, of existing possibilities of technological and organisational innovations on lines and principles established before but steadily improved in the process; revision of the whole structure of industry in quest of increased efficiency; systematic struggle with each item of the list of costs ... (Schumpeter, 1939, 759)

This is clearly a situation for managers but not for gambling entrepreneurs.

The above-mentioned biologists MacArthur and Wilson (1967; Pianka, 1983) have developed a somewhat similar discussion which led to the distinction between r-selection and K-selection as well as species which are r-strategists (i.e., pioneering-type behaviour) and K-strategists (i.e., crowding-type behaviour). In the case of biological species we are dealing metaphorically with the term 'strategy' which relates to a behaviour which is genetically determined. The idea might be even more relevant to the area of economic evolution where the strategy concept is more obviously adequate. The more systematic view proposed by the ideas of MacArthur and Wilson seems to clarify some of the limitations of Schumpeter's discussion of the railway case. We may say that he overemphasises r-selection and the related innovative-entrepreneurial behaviour while he is to some extent neglecting K-selection and the related managerial behaviour. He ignores the extent that he sometimes denies it any role in a long-term evolutionary perspective. A little reflection will suggest a significant evolutionary role for the K-strategists who may even produce the bulk of evolutionary change. At least it is clear that it is important to have an analytical framework which is broad enough to cope systematically with both kinds of selection and both kinds of strategy.

The difference between Schumpeterian entrepreneurs and managers may be related to the economic history of the United States. The 'frontier' was the area of a simple type of settlement and entrepreneurship while established regions tended to become crowded with respect to these simple forms of business. However, if an individual accepted the tight competition and the more regular mode of behaviour needed when everything settles down into a kind of quasi-equilibrium, then he could make a living under more stable conditions. Later the idea of the frontier was generalised, and science and innovative business became viewed as kinds of 'endless frontiers' where there were always possibilities of pioneering or entrepreneurial behaviour. In this respect the creation of a new combination or new paradigm means the opening up of a new frontier area which, nevertheless, will be crowded and not particularly profitable. To survive, it is necessary to show a highly adaptive behaviour as well as to avoid the vulnerability involved in risk-taking. Some do not want to survive under such conditions and try to escape by creating yet another frontier.

We cannot explore the interesting differences between pioneering and crowding modes of selection and of strategy. However, we should remark that they emphasise an important result. In our evolutionary analysis the starting point is not the actors but the routines that they apply.[29] If the frequency of the routine increases, we may say that it shows a supernormal fitness in the struggle for economic life. This kind of argument has been severely criticised by Popper (1972). Nevertheless, in the biological and philosophical counterattack it has been argued convincingly that the evolutionary arguments may be cleared up if they take into account the crucial distinction between theoretical and observed (tautological) fitness (see also Winter, 1987, 615). In the present interpretation another crucial distinction has emerged. The problem is that both the theorist and the economic decision-maker have a theory about the fitness of a given routine. Therefore, we have three notions of fitness:

1. Theoretical fitness is the theorist's more or less grounded expectations about the changing weight of the economic activities based on a routine type.
2. *Ex ante* fitness is the economic decision-maker's expectation about the fitness value of a certain economic routine.
3. Observed or *ex post* fitness is just another name for the observed change in the frequency of the economic applications of a routine.

The speculation about *ex ante* fitness is a central aspect of the activities of innovative entrepreneurs. They have no place in models which do not allow this area to be a question of judgement. However, entrepreneurship is also a question of implementation. Therefore, we may also say that innovative entrepreneurs have no place in models and economic situations which make the implementation of all investment projects a simple matter.

Notes

1 The critiques are indirectly reflected in Schumpeter (1939, 205, 492).
2 The present chapter has to focus on certain aspects of this literature. Quite another chapter could have been written if the focus had been placed on the approach to complex systems developed within far-from-equilibrium thermodynamics (Allen, 1981; 1988). However, such a chapter would be much more abstract and thus increase the distance to the analysis of the network of interactions between the application of routines.
3 The unrealistic conception of the search space led to serious misconceptions of the policy implications of the growth pole theory, both in regions and in developing countries. To put the problem sharply: the experience of interdependent and dynamic development based in the steel production of the Ruhr area in the nineteenth century (which was Perroux's paradigmatic example) did not help in providing rapid development in the developing countries where it has been applied. Neither did the fact that steel production shows large multipliers in the input-output tables of developed economies. The problem of being more than a hundred years too late is not necessarily that there do not emerge as many innovative ideas and possibilities from steel production as before, but rather that the 'economic niches' suggested for the suppliers and customers of a steel mill are already occupied by well-established firms/industries. Furthermore, there are well-established relationships of the innovative type in the developed countries which are more likely to give ideas for

new 'niches'. Finally, the industry has changed character, thereby, perhaps, making non-formalised contacts less important and easier to establish across national borders than before.

4 A somewhat naive presentation of the state of the art in the 1920s is found in Kuznets (1930/67, 64-69, 291), where the logistic as well as the Gomperz curve is treated. The empirical use of the logistic curve, especially in demographic analysis and prognosis, created an overheated debate in the 1920s and 1930s. The major proponent for the use of the model was the biologist Pearl of Johns Hopkins University while the well-known statistician E.B. Wilson and other staff of Harvard University (including the department of economics to which Schumpeter moved in 1932) were astonishingly sharp in their critiques (see Kingsland, 1985, chs. 3-4). In such an environment Schumpeter's (1933/51) own naive idea of an alliance between economic theory and statistical and historical analysis (the econometric research programme) was in great trouble.

5 See popular introductions like Hofstadter (1985a) and Gleick (1988).

6 The program PHASER described by Koçak (1986).

7 The names of the parameters are taken from the normal notation in ecological modelling. The reason for this is that this ecological notation is wide-spread while there is no standard economic notation. An example of a very shifting notation can be found in Goodwin, 1990a. In economics it is important to emphasise that K is not capital but the maximum sustainable level of the application of an innovation. The analogy to the biological 'saturation level' is not perfect since K is not normally a constant of the diffusion process, see below. Another problem is that the logistic difference equation may be formulated in other ways than equation 3.1. Goodwin (1990a, 14) shifts between two other formulations. The most abstract formulation is based on the transformation of the variable into the size of the 'population' relative to its maximum sustainable size. In both cases Goodwin's formulations can be transformed into the present form by a recalculation of the parameters.

8 The last formulation is the one which is further developed in the Lotka-Volterra model, see equations 3.5.

9 For high values of r we find stable cycles around K and for even higher values ($r >$ 2.570) we find deterministic chaos which was actually discovered by studying the logistic difference equation (see May, 1976/81, ch. 2). This level of r has perhaps some relevance in the development of decision-making during 'railway manias' but not when it comes to real length of railway lines. But it must be emphasised that the logistic difference equation is a purely formal tool and we cannot be sure that it is possible to give it any precise economic meaning.

10 The discussion here is made in terms of difference equations rather than differential equations while Goodwin applies a mix of them. In the present context the discrete formulation is a convenient mode of exposition. Further arguments are: First, evolutionary process embraces both discrete and continuous aspects of economic life (just like the digital code of the egg and the analogue mode of behaviour of the hen). The basic problem of contemporary evolutionary-economic analysis is, however, to include explicit microfoundations. This means that we cannot always remove the discrete formulations by turning to a higher level of aggregation. Instead we should make it relatively easy to introduce innovations which are basically discrete events. The continuous formulation of the diffusion of the innovation will first be relevant when we turn to an innovation which has entered its 'paradigmatic' phase and has reached a certain degree of application. Second, the discrete formulation makes clear the way we in simulation studies are manipulating the results by means of a more or less complex lag structure. Third, the discrete formulations help us to remember the difficulties of a computer-implementable concept of diffusion and evolution: the computer is basically applying discrete steps of calculation and any smooth curve is the result of making the time steps become smaller and smaller.

11 Schumpeter would probably argue that the latter will also be forced into shorter cycles because the reaction mechanisms of the economic system tend to make decision-making about large innovative investment projects develop in a stop-go

manner while the corresponding creation of capital has a more smoothly logistic form because of the inertia of the investment process. This part of Schumpeter's argument will, however, complicate the modelling work and is (probably for this reason) ignored by Goodwin. Instead he assumes that we may operate with the short and long logistic curves independently of each other.

12 The question is not what happens to the short cycle when it is affected by the long cycle (and *vice versa*). By answering this question we may be able to rethink Schumpeter's well-known scheme of superimposed cycles (Schumpeter, 1939, 213) in a way which does not simply add two types of given oscillations, or even three types of sinus-waves considered as representing Kitchin cycles, Juglar cycles and Kondratiev waves. At present it is, however, only the basic idea which shall be presented.

13 In the attempts to remake Goodwin's poorly documented calculations, a programming package for continuous modelling was used, namely the Macintosh version (STELLA/ITHINK) of DYNAMO, the programming language and programming 'philosophy' for Forrester's Systems Dynamics. Discrete elements were introduced into the calculations by means of changing the stocks in a step-wise manner (by dividing with the time-step). However, the difficulties in remaking calculations (compare figure 3.4 with Goodwin, 1990a, fig. 4.2) reflect not only the present simulation package but also the poor documentation of Goodwin's work.

14 For simplicity a complete adjustment between output and demand is assumed to be obtained during one period.

15 In his formulation of his version of a similar expression Goodwin (1990a, 44) is imprecise. The present version is an interpretation.

16 This assignment statement includes several changes *vis-à-vis* Goodwin's (1990a, 44) formulation.

17 A specialised program for this exercise is recorded in the following program which is a revised version of an original made by Jørgen Østergaard, University of Aalborg.
Program Goodwin 3.4.discrete
memory1(t) = memory1(t - dt) + (Qt2 - Qt1) * dt
    INIT: memory1 = Qstart*stepsize
    INFLOW: Qt2 = a*Qt1 - b*Qt + delta + kappa*It1
    OUTFLOW: Qt1 = memory1/DT
memory2(t) = memory2(t - dt) + (Qt1 - Qt) * dt
    INIT: memory2 = Qstart*stepsize
    INFLOW: Qt1 = memory1/DT
    OUTFLOW:: Qt = memory2/DT
Nt12(t) = Nt12(t - dt) + (It1) * dt
    INIT: Nt12 = Nstart*stepsize
    INFLOW: It1 = r*Nt12*(1 - Nt12/Klong) + c*Qt1
a = ?
b = ?
c = ?
delta = ?
kappa = ?
Klong = ?
Nstart = ?
Qstart = ?
r = ?
stepsize = DT

18 Goodwin offers the analogy by the so-called Roman fountain, which fills to a certain point before a rapid outflow occurs. This analogy and this subsequent presentation is not particularly illuminating, see Goodwin (1990a, 42).

19 It is not my intention to discuss biological evolution but to try to develop the biological analogy systematically. Furthermore, the disanalogies are just as important

as the analogies in order to avoid some of the confusion which surrounds many evolutionary discussions in social science.
20 Or rather: they think they foresee the values around which the prices may oscillate.
21 Due to the time-lags of economic decision-making and of the investment process, we may often have to introduce period $t-1$ and even earlier periods. In the following we stick to the simple formulations.
22 Which may be generalised into a system of equations for n interacting routine types. The A-matrix can then be studied qualitatively.
23 The different types of ecological interactions are presented by Boulding (1978, ch. 4). However, Boulding discusses biological populations and does not use the scheme for describing economic affairs.
24 A simple discussion of some of the intricacies involved can be found in Haberman (1977).
25 4 ft 8 1/2 ins.
26 The development of the concept of a niche is reflected by the papers in Whittaker and Levin (1975). See also Pianka (1983) and Hannan and Freeman (1989, ch. 5).
27 By a close examination of this set of equations, we may develop a discussion like the previous one. A more complex discussion about how a firm finds a suitable niche may also be included (see Lloyd et al., 1975, 127-133). For the present exposition another strategy has, however, been chosen. We turn to the case of many firms and consider some aspects of the decision-making of a single firm.
28 A similar emphasis on judgmental issues as a *differentia specifica* of the entrepreneur is found in Casson (1982).
29 An alternative, individualistic approach is, e.g., developed by Witt (1987).

4. Exploring the process of 'Schumpeterian competition'

4.1. The complex world

4.1.1. Introduction

Nelson and Winter[1] were probably the first to create computer programs and calculations which with some right can be interpreted as reflecting the mechanisms of an evolutionary-economic process. Their computations and underlying models deal with an industry or an economy with several firms whose behaviour is based on different kinds of routines or rules.[2] These rules are similar to the ones explored by the behavioural school of economics (Simon, 1982; Earl, 1988). In a concrete model and the related computer simulations of it, some of the rule-types are taken as parameters while other rule-types are allowed to evolve. The evolution of the latter is based in a process of 'Schumpeterian competition' where firms introduce and imitate rule-variants in order to succeed in a market game which goes on for many periods. This is the basic idea of Nelson and Winter's very preliminary version of what is here called Artificial Economic Evolution.

Nelson and Winter have primarily promoted their idea of an evolutionary process by means of their concrete models and simulation results. But this part of their work is not what is of most interest in the present context. When dealing with the relationship between new evolutionary economics and the work of Schumpeter, a much more important aspect of Nelson and Winter's work is that they have created the outlines of a language and a modelling scheme which help us express the poorly understood evolutionary processes. It may thus raise a series of questions which may lead to an effort to increase our understanding of these processes. It is not the purpose of this chapter to study whether the Nelson-and-Winter models reflect what Schumpeter means by the evolutionary role of the competitive process. The purpose of the chapter is instead to present and discuss a possible mode of articulating 'Schumpeterian competition'.

The following exposition of the contribution of Nelson and Winter will concentrate on their explicit and implicit language and modelling scheme rather than on their many simulation exercises. In contrast to the situation faced by the old evolutionary economics, we now have the ability to express in detail many parts of an evolutionary process. In dealing with these apparently unnecessary details, we will prepare ourselves for a rethinking of Schumpeter's work but also for judging to what extent Schumpeter still represents a challenge to the new evolutionary economics. In section 4.2 we discuss the character and main elements of Nelson and Winter's evolutionary synthesis and present the outlines of a typical Nelson-and-Winter-model of 'Schumpeterian competition'. This model is expressed more formally in section 4.3, and in section 4.4 several extensions of it are introduced (satisficing behaviour, entry and exit of firms, industry creation, product change). Finally, we turn to the broader issues of a computer-based evolutionary economics in section 4.5. However, first of all we have to consider the background of rule-based behaviour.

4.1.2. Problem complexity and Artificial Intelligence

One of the most important general contributions of the twin areas of computer science and mathematics of computation is that they have provided an increased understanding and ability to cope with different kinds of complexity,[3] issues which are also of major importance for evolutionary analysis (see, e.g., section 5.3.3). The contributions to the understanding of complexity are developed because in the computer world we find something which does not exist in the real world (except in relation to highly stylised problems), namely 'an infallible step-by-step recipe for obtaining a prespecified result.' (Haugeland, 1985/89, 65) Since a given task may be solved by different algorithms, we have to evaluate and select algorithms in terms of different performance measures (Knuth, 1968/73, ch. 1): Is it really an algorithm which solves the task for any allowed input? Is it an ineffective algorithm whose use of resources (time and memory) increases as an exponential function of the size of its task or is it able to solve larger tasks within realistic limits? Do there exist better algorithms for solving the same task?

The analysis of algorithms is or might be influencing new evolutionary economics in several ways. First of all, we may consider the heterogeneous behaviour of the economic agents as based on different algorithms or, rather, heuristics (see below). In this perspective the evolutionary process is one of inventing and testing different algorithms or heuristics. Second, we now have programming languages which make it easier for the researcher to express a given algorithm clearly. Third, the complex mechanisms of interaction between the different agents of an evolutionary process may be modelled by means of algorithms which help the researcher to cope with the system-level complexities.[4] Fourth, the clear articulation of the algorithmic or computational approach has helped to emphasise the limit of the Walrasian approach with a clarity which was unthinkable for many of the contributors to old evolutionary economics.

The complexity of the problems which practical computing sometimes tries to deal with appears to demonstrate that many Turing-computable algorithms cannot be computed within realistic resource limits (see, e.g., Lewis and Papadimitriou, 1981, ch. 7; Rayward-Smith, 1986, ch. 6). To make this question more precise we define the number of computational steps of an algorithm or a Turing machine as a function of the input. For example, we may study the famous travelling salesman problem which deals with a salesman who wants to visit all the cities of his sales district in a way which minimises his costs of travelling and which starts and ends in his home city (Karp, 1986; Lawler *et al.*, 1985). To simplify we may allow cross-country travelling; in this case the number of possible tours for the travelling salesman in a district with n cities is a very rapidly growing function of n with $(n - 1)!/2$ possibilities, a number that rapidly gets astronomical and hyper-astronomical as n increases: we have a case of combinatorial explosion. Karp (1986, 99) has calculated that already for $n = 20$, a study of all possibilities takes more than a thousand years even if we study one tour per microsecond.

But we do not always have to check all possibilities. Other algorithms have been invented which perform much better even if they still show a tendency of exponential growth. That much can be done by better algorithms is demonstrated through the invention of methods able to solve the travelling salesman problem for n up to more than 300. Before this enormous improvement by Karp and others, the only alternative for $n > 16$ was to give up the search for a globally optimal solution and to use heuristic (normally local) search for finding 'satisfactory' solutions. Are we now approaching a situation where we can transform exponentially hard solution methods (showing combinatorial explosion) with polynomial time solutions which are of practical use? Unfortunately, Cook, Karp and others have proven that our inventiveness will never avoid computational explosion as n continues to rise, unless *all* members of a huge class of difficult optimisation problems of economics and engineering (where solutions are difficult to find but easy to test) find a non-exponential solution. Even if the ill-behaved character of this interesting class of problems has not been mathematically proven, we are next-to certain about this character. The reason is that mankind has always been confronted with this kind of problem and has not yet come up with a non-exponential solution of the optimising type.

Artificial Intelligence[5] is to a large extent concerned with complex problems for which it is impossible to find globally optimal solutions. Actually, this was the original problem for the main founders of Artificial Intelligence: Newell, Shaw and Simon. For them problems are exponentially hard problems and human intelligence is the ability to solve such problems in an approximate way:

We have begun to learn how to use computers to solve problems, where we do not have systematic and efficient computational algorithms. And we know, at least in a limited area, not only how to program computers to perform such problem-solving activities successfully; we know also how to program computers to *learn* to do these things. [/] In short, we now have the elements of a theory of heuristic (as contrasted with algorithmic) problem solving; and we can use this theory both to understand human heuristic processes and to simulate such processes with digital computers. Intuition, insight, and learning are no longer exclusive possessions of humans: any large high-speed computer can be programmed to exhibit them also. (Simon and Newell, 1958, 6)

These formulations are programmatic and overoptimistic but they reflect much of the subsequent work in Artificial Intelligence: AI = high-speed computers + heuristic problem solving. However, it should be emphasised that just as computers had a non-electronic prehistory from the 1930s, so the theories of (hard) problem solving had developed in the same period. One of the connecting links is Simon[6] who had developed his core ideas of the conditions and character of administrative and economic problem solving and behaviour (e.g. Simon, 1947/65) before he (in the mid-1950s) also became a cognitive and computer scientist mainly specialising in Artificial Intelligence; but the further development of his ideas of bounded and procedural rationality (Simon, 1982) is clearly influenced and supported by his experiences in programming and computer science. Thus, when he is asked for substantial theories to guide empirical inquiry in behavioural economics, he points to his Artificial Intelligence work on complex problem-solving situations (Simon, 1986).

The background for rule-following and meta-rule-following behaviour is found in heuristic decision principles for coping with problem complexity. In Greek *heuriskein* means 'to discover' and as an adjective 'heuristic' means 'serving to discover'. As a noun a 'heuristic' denotes the 'study of methods and rules of discovery and invention' (Polya, 1945/71, 112) as well as an individual instance of a principle of discovery. There are no generally secure methods of discovery and therefore heuristic principles are just guides which help to improve the efficiency of search processes. When we as tourists search a foreign city we may use a book or a person as a tourist guide; in this way we avoid a lot of random search but also possible discoveries and amusements. If we use good a heuristic principle (a heuristic), we may hope to achieve good but not globally optimal solutions to highly complex problems. Even in a formalised version heuristics are not algorithms (which are defined to be deterministic and infallible) but rules of thumb. But even heuristics may be formulated in terms of search routines for which we are often able to find an algorithm! However, it is an algorithm which in a well-defined way finds a satisfactory solution to the goal set for guiding heuristic search, not an optimal solution to the original complex problem (Haugeland, 1985/89, 82-84). Thus, it does not support the kind of decision-making depicted by the tradition of a *substantively* rational decision-making[7] where the choosing among alternatives involves the following series of steps (see Simon, 1947/65, 67):

1. Listing all the alternative strategies.
2. Determining all the consequences that follow upon each of these strategies.
3. Comparative evaluating of these sets of consequences based on fully known preferences.

4. Implementing the decided strategy in full accordance with the plans.[8]

To Simon, the sharp articulation of these characteristics of rational economic man (*Homo rationalis* or *Homo economicus*) is only a means of studying the limitations of this creature and gaining insights about the real 'administrative man': 'It is obviously impossible for the individual to know *all* his alternatives or *all* their consequences, and this impossibility is a very important departure of actual behaviour from the model of objective rationality.' (Simon, 1947/65, 67) In real life it is impossible for a decision-maker to reach any high degree of substantive rationality since the 'number of alternatives he must explore is so great, the information he would need to evaluate them so vast that even an approximation to objective rationality is hard to conceive.' (p. 79) Instead he has to accept 'bounded rationality' (Simon, 1982) as a *condition humaine*. This means that 'only a very few of all these possible alternatives ever come to mind', that 'knowledge of consequences is always fragmentary', and that 'values can only be imperfectly anticipated' (Simon, 1947/65, 81) As a consequence, the process of 'implementation' becomes to a large degree a discovery process[9]. The development of the theory of computational complexity has for vast classes of problems proved these statements and emphasised the need of analysing alternative ways of obtaining a realistic rational behaviour.

Simon's argument for the dominance of boundedly rational behaviour leads him to suggest the ubiquity of 'procedural rationality' in which the decision-maker is occupied with retrospection rather than real foresight.[10] The main form of procedural rationality studied by Simon is found in his model of satisficing behaviour where the decision-maker 1) considers the major part of reality as consisting 'of "givens"—premises that are accepted by the subject as bases for his choice' (p. 79), and 2) attempts to challenge the 'givens' and search for alternatives only when his performance becomes unsatisfactory. The model of satisficing behaviour is based on two types of heuristic principles. 1) If you have found a satisfactory behavioural rule, you should stick to it unless performance falls below a critical value which is given by experience. Or: Same procedure as last year, unless this becomes intolerable. 2) When you have started a search for an alternative rule, you should know when to stop. If you are searching for needles in a haystack, you should beforehand decide to stop if you find a not-too-rusty needle (even if you may soon find a better if you proceed) or if you have spoiled a whole day's work. In other words, you should not only have a search procedure but also a test which decides whether you have found a satisfactory solution, and you may also have other ways of avoiding indefinite search (Newell and Simon, 1976). The heuristic principles may also deal with the problem of how to search for new rules in a non-random way, e.g., through localised search, etc.

Much of the research in Artificial Intelligence can be seen as attempts to explore highly complex decision situations from the decision-maker's perspective and to express and explore important decision rules.[11] The key elements of this approach to the simulation of problem solving is:[12]

1. Define the problem situation with different potential states of affairs, operations by which the decision-maker can change the state, and goals by which he can evaluate alternatives.
2. Make a computer representation of the problem situation.
3. Make a computer simulation of the search process among alternative courses of action in order to find those that will lead to the desired goal to a satisficing degree.
4. Chose the alternative which best achieves the desired goal.

The major difference between this scheme and the rationalistic paradigm is the attempt to make an explicit study of the search process in the problem space (or, rather, its computer representation). At the same time, it must be emphasised that for Simon these studies have led to computer science, psychology and their combination in Artificial Intelligence rather than to the economic and broader social set-ups within which they were originally conceived. For this reason there has been little development of his famous considerations about bounded and procedural rationality and his other contributions to economics (Simon, 1991a, ch. 21) [13] The application of his ideas has been left to their researchers with less knowledge of complexity theory and Artificial Intelligence.

Cyert and March are two of Simon's colleagues who tried to develop the ideas of rule-following behaviour in *A Behavioural Theory of the Firm* (1963). Here we find detailed models and detailed computer simulations of decision-making in single firms. In this way much organisational realism is added to Simon's schemes and even an initial testing of the satisficing hypothesis is performed (*ibid.*, ch. 7). The latter is formulated as a rather successful attempt to compare the actual and the computer simulated (based on satisficing) pricing-decisions in a large retail store. The book and similar works of the Behavioural School are now mainly remembered for their organisational arguments for rule governed behaviour: the firm is an adaptive, imperfectly rational coalition of different interest groups whose bargaining processes lead to rules which may be considered as a kind of truce in interorganisational conflict. Such a firm may have considerable organisational slack which absorbs a substantial share of the environmental shocks while the given rules are upheld. Only when performance does not live up to the acceptable-level goals, is a 'problemistic' search for alternative rules then performed.

We may, in the work of the Behavioural School. see the beginning of a theory of developmental processes which takes place in an interplay between the different parties involved in a given firm. Another contribution to such a firm-oriented theory is found in Penrose's *The Theory of the Growth of the Firm* (1959/80). From the viewpoint of the present book such studies are concerned with ontogenetic evolution (developmental processes of individuals) rather than the kind of evolution with which we are mainly concerned, namely phylogenetic evolution or evolution of the species, which must include an explicit population level.[14] Both types of studies are, of course, important, but there is a trade-off. If we focus on the detailed behaviour of the individual decision-maker or the complex firm, as Simon and other behaviouralists do, then attention is drawn away from the overall changes at the population level and the competitive processes which drive these changes. However, their studies contribute to the understanding of problem complexity as part of the 'origin of predictable behaviour' (Heiner, 1983; 1989) and this contribution is a necessary precondition for the development of the new evolutionary economics.

4.2. The pioneering work of Nelson and Winter

4.2.1. An evolutionary synthesis

Nelson and Winter presented their versions and interpretations of what I call Artificial Economic Evolution in a series of papers (from Nelson and Winter, 1973; 1974 to Winter, 1984/91) and in the related book, *An Evolutionary Theory of Economic Change* (1982). Through this work they demonstrated the possibility of overcoming the basic difficulty in studying evolutionary processes, namely the need to combine elements in a complex manner which are normally considered as belonging to quite different areas of investigation. Such a combination presupposes two opposing qualifications: an ability to cope with a wide diversity of elements and an ability to cut out the details and integrate the elements into an initially crude conception of an evolutionary process. Only in this way is it possible to create a framework which makes sense of the processes of transmission and variety-creation. The computer helps to organise this exercise including the last steps of synthesis since 'the simulation format does impose its own constructive discipline in the modeling of dynamic systems: the program must contain a complete specification of how the system state at $t + 1$ depends on that at t and on exogenous factors, or it will not run.' (Nelson and Winter, 1982, 208 f.)

The evolutionary models created through the collaboration of Nelson and Winter build on their previous individual scientific works. Before their joint work, Winter (1964; 1971/93) had already made important critiques of the Alchian—Friedman selection argument for profit maximisation (see section 1.1.3) while Nelson had been working on the economics of invention, innovation and technical change (Nelson, 1959a; 1959b; 1968). Through their common research endeavour they made an evolutionary synthesis by integrating ideas about:

1. Behavioural patterns and their transmission.
2. Creation of new behavioural patterns.
3. Different types of selection mechanisms.

More specifically, we may say that they combined:

1. Simon's work on rules and satisficing behaviour.
2. Nelson's and other 'Schumpeterian' work on invention and innovation.
3. Alchian's and Winter's work on 'natural selection'.[15]

In the creation of the evolutionary synthesis there is little doubt that the contribution of Simon (and Cyert and March) was of crucial importance.[16] However, the evolutionary synthesis is a clear example that the whole is more than its constituent parts. Furthermore, we see that the elements were reshaped to fit into their new place in the evolutionary synthesis. Therefore, it is appropriate to present the models as if they were wholly created through Nelson and Winter's joint work even if they are heavily indebted to several sources.

Nelson and Winter's evolutionary models are based on the postulate that it is possible to specify the space in which innovative search takes place as well as the way the actual search process takes place. In other words, we postulate a degree of stochastic predictability of most innovative activities which may allow the formulation of 'laws of succession', of course in probabilistic terms and subject to *ceteris paribus* clauses. Behind the 'laws' are firms' and financial organisations' rules of searching and decision-making which are in many ways bound to their present state. The search for new rules often starts with problems arising from existing rules, and the result of the search will be evaluated by comparison with these rules. This predictable aspect of economic change may be seen as a result of bounded rationality leading to localised search in the space of (technological and marketing) alternatives. Here distance is measured in terms of search costs, and there is increasing uncertainty about the precise characteristics of alternatives. What the firm applies as a relatively stable solution will, at best, express a temporary local optimum.

Generally, we try to imagine the state of a firm at period t with respect to, e.g., physical capital, productivity, product characteristics (in industries with heterogeneous products), etc. Together with output rules and functions of factor supply and final demand, this state determines the firm's competitiveness *vis-à-vis* its similarly described competitors and thus its profitability in period t. The firm's state in the next period, $t + 1$, is determined by its (simplified) investment rules and by its search rules (and thus search costs) together with the (assumed) probabilities of finding new rules in the space of alternatives. A newly found alternative will only be included into the new state of the firm if it is judged to increase expected profitability.

This modelling strategy may be summarised in the following way:

1. Define the minimum environmental characteristics, including input and output conditions as well as the spaces in which search for new rules is performed.
2. Define the state of the industry at time t as a list of firm states which include physical and informational characteristics as well as behavioural rules and meta-rules.
3. Calculate by means of (1) and (2) the activities of the industry in period t as well as the resultant state variables (including possible changes of rules) which characterise the system at the start of period $t + 1$.
4. Make similar calculations for a series of periods and study the evolution of the application of different rules as well as other characteristics of the industry (economy).

By accepting such kinds of elements in their model-makers' tool-kit, Nelson and Winter are imposing upon themselves a certain conception of the evolutionary process. First of all, they apply a population perspective. An 'industry' or an 'economy' is seen as a taxonomic class incorporating a certain degree of variety of processes and/or products; but the variants must, in principle, be transferable between the different firms. This also implies a certain similarity of the search spaces of the firms, although there may be major differences with respect to the 'distance' to different sources of knowledge. The reader should be aware that the empirical relevance of the whole argument is heavily dependent upon the possibility of defining a level of aggregation and a related taxonomy which are not arbitrary constructs of (national) statistical services but which instead reflect important similarities and differences with respect to the factors of the evolutionary process. Second, the name of the game is variety-creation and variety-selection within a given economic pattern. In other words, Nelson and Winter emphasise change which follows 'natural trajectories' within given 'technological regimes' (Nelson and Winter, 1982, 258-262) or 'technological paradigms' (Dosi, 1982) rather than radical change. The latter is suggested in contrast to their models rather than within their models. Third, 'a vast array of particular models can be constructed within the broad limits of the theoretical schema' but the 'enormous generality' of the schema cannot be exploited immediately (Nelson and Winter, 1982, 19 f.). To obtain real understanding about how to handle their powerful family of models, Nelson and Winter prefer to concentrate on 'very simple examples' and to 'distinguish sharply between the power and generality of the theoretical ideas we employ and the much more limited results that our specific efforts have yielded thus far.' (pp. 20 f.)

4.2.2. Typical Nelson-and-Winter simulation models

In their attempts to exploit their highly general modelling scheme, Nelson and Winter have developed several generic and specific types of models.[17] To grasp their basic ideas, we have to concentrate on one central example. It is their most developed and documented model type (see Nelson and Winter, 1982, chs. 9, 12; Winter, 1984/91) which deals with the evolution of the production techniques and other behavioural rules of an industry producing a homogeneous product. This species of model may be considered as Nelson and Winter's major contribution to Artificial Economic Evolution and it includes the elements of decision rules, search and selection.[18]

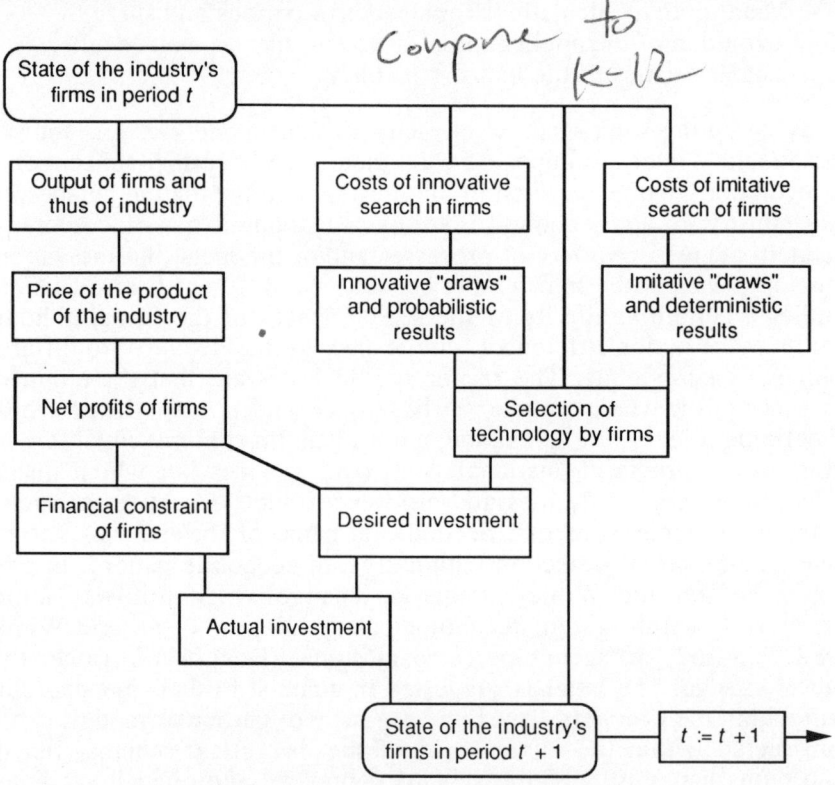

Figure 4.1. The computational structure[19] of simple simulation models of Nelson and Winter.[20]

By concentrating on this model type we also have the opportunity of applying some techniques like flow charts and an algorithmically oriented notation (see the appendix on the latter issue). These techniques may be more helpful in reflecting upon simulation models than the usual kind of mathematical notation. The latter method of exposition is used by Nelson and Winter and practically everybody else. However, we occasionally see an uncontrolled application of computerese which illustrates the problem (see Nelson and Winter, 1978, 544-547).

Figure 4.1 describes the computational structure of the Nelson-and-Winter model variant on which we shall concentrate in the following. Like all the other Nelson-and-Winter models, it determines (probabilistically) what happens in each period. This period has inherited a state of the industry from the former period. The state is defined in terms of the size of the physical capital stock and the productivity of capital of each firm.

The computational steps of the figure describe how the state of the industry in the next period is found when the state (capital stock and productivity) of the present period is given.

First, the left column of the figure shows a simple short-run system, i.e., a simplified economic process in the industry whereby output, price, profits and financial constraints of firms are found.

1. Output of each firm is decided by simple capacity utilisation *rules* of firms, e.g., full capacity utilisation.
2. Output of the industry is found by simple aggregation.
3. The aggregate output of the industry faces exogenously given demand conditions, e.g., a conventional downward-sloping market demand curve. This gives the market-clearing price per unit of output.
4. For each firm we calculate the turnover and then find the net profit by deducting capital depreciation, variable production costs and search costs (which are, e.g., all given per unit of physical capital).

Second, the upper parts of the middle and right columns of figure 4.1 show the processes whereby new production techniques are found and productivity is changed. In the particular model under consideration firms are always involved in a search for better production techniques. Because of this, they explore a space of possible production routines which are defined in terms of capital productivity.[21] Their search costs are defined by fixed relationships to their capital stock and are of two types: innovative search and imitative search. We shall start with the simpler case of imitation (even if innovation is logically prior to imitation).

1. The effort and costs to imitate the techniques of other firms are given by the firms' *rules* of determining the size of the budget of imitative search (which may also be classified as imitative R&D). In the present Nelson-and-Winter model the rule is the same in all firms, namely to use a fixed amount of imitative search costs per unit of capital.
2. Because of its imitative search effort, each firm gets access to a 'lottery'. Its probability of obtaining a 'draw', i.e. to draw a ticket from the lottery,[22] is proportionate to its imitative search costs but

is otherwise determined by exogenous factors (the difficulty of imitation in the particular industry).
3. A 'draw' means that the firm gets access to the best-practice technique and thus the highest productivity level obtained by any firm in the period.
4. The firm's *rule* concerning the costs of innovative search (innovative R&D) is also to spend a fixed amount on innovative search per unit of capital.
5. The firm's chance of getting a 'draw' (a successful innovation) in the innovative 'lottery' is proportionate to its innovative costs as well as to the exogenously given character of technical change in the industry. The technique found by a 'draw' depends on exogenously given probability distributions. To give an impression of such distributions we may consider two major types (Nelson and Winter, 1982, 211 f., 283). The simplest case is a 'science based' industry where the probability distribution is only dependent on exogenous factors. In the case of an industry characterised by production-near 'learning processes' and thus 'cumulative technology', the value of a 'draw' is most likely to be near the firm's present technology (localised search).
6. The attempts to improve productivity end with a comparison between the productivities obtainable by the technique inherited from the last period and the techniques which may be found by imitative and innovative search. The technique with the highest productivity is chosen.
7. If the technique is changed, it will determine productivity of the next period (disembodied technical change). We thus have the state of technique (production routines) for period $t + 1$.

Third, the lower parts of figure 4.1 are concerned with the investment decision.

1. The only way to reduce productive capacity is through the process of physical depreciation, but this is not depicted by the figure.
2. The firm's gross investment function is bounded below by the non-negativity condition and above by the financial constraint. The financial constraint is determined by the net profit which is increased by external financing in some ratio to net profit.[23] This allows a primitive treatment of the role of banks' rules in the evolutionary process (see Nelson and Winter, 1982, 291 ff.).
3. The firm's desired net investment depends on a) the ratio of the price of the period to unit costs of the next period (taking into account productivity change), b) a target mark-up factor which is an increasing function of the firm's market share. The desired net investment responds positively to the price/cost ratio and negatively to the market share.

4. Actual investment depends, of course, on desired investment as well as investment constraints.
5. The investment process has no time-lags. The adjusted physical capital stock is available to the industry's firms in period $t + 1$. By multiplying the capital stock with the new level of productivity, we have the production capacity of the firms of the industry in period $t + 1$.

And so the computation goes on and on for $t + 1, t + 2, ...$

This exposition of the basic computational structure of the typical Nelson-and-Winter model does not specify fully the simulation programs. On the contrary, a large number of decisions has to be made before we have a functioning computer program and/or a full mathematical specification of the Markov chain. However, the exposition has more or less followed the rules of structured programming (Dahl *et al.*, 1972) and may later open up for modular programming (Wirth, 1982).[24] This means than we may apply a top-down approach to programming where the computation is gradually decomposed in more and more detailed tasks. Furthermore, it means that most of the 'boxes' in figure 4.1 may to some extent function as modules with relatively simple interfaces *vis-à-vis* other 'boxes'. This means that each module can be specified in relative isolation from the other modules. More importantly: a primitive version of a module can later be exchanged with a more sophisticated version; an old version of a module which reflects one hypothesis can be exchanged with a new version which reflects another hypothesis.

Figure 4.2 (see next page) serves to emphasise this structured and more or less modular design as well as to depict the structure of the above comments on figure 4.1. In the new figure we emphasise the need to make a procedure INITIALISE which defines the data structures for the process of computation. This does not only include the state variables (the initial capital stock and productivity of each of the n firms of the industry) but also the parameters which, e.g., specify the behavioural response of the firms and the demand side of the market. The SHORTRUN procedure defines the market process based on given behavioural routines, given capital stocks and productivities, and a specification of the demand side of the market. A small set of procedures (INNOVATION, IMITATION and TECHNCHOICE) defines how the firms try to exploit the technological search space as well as each other's capabilities and determines the resultant productivity of each of the firms. Similarly, a small set of procedures (MAXINVEST, DESIREDINVEST and REVCAPITAL) defines the investment and depreciation process of firms. Taken together the procedures deliver inputs to the master procedure UPDATE which modifies the state variables after each period of computations.

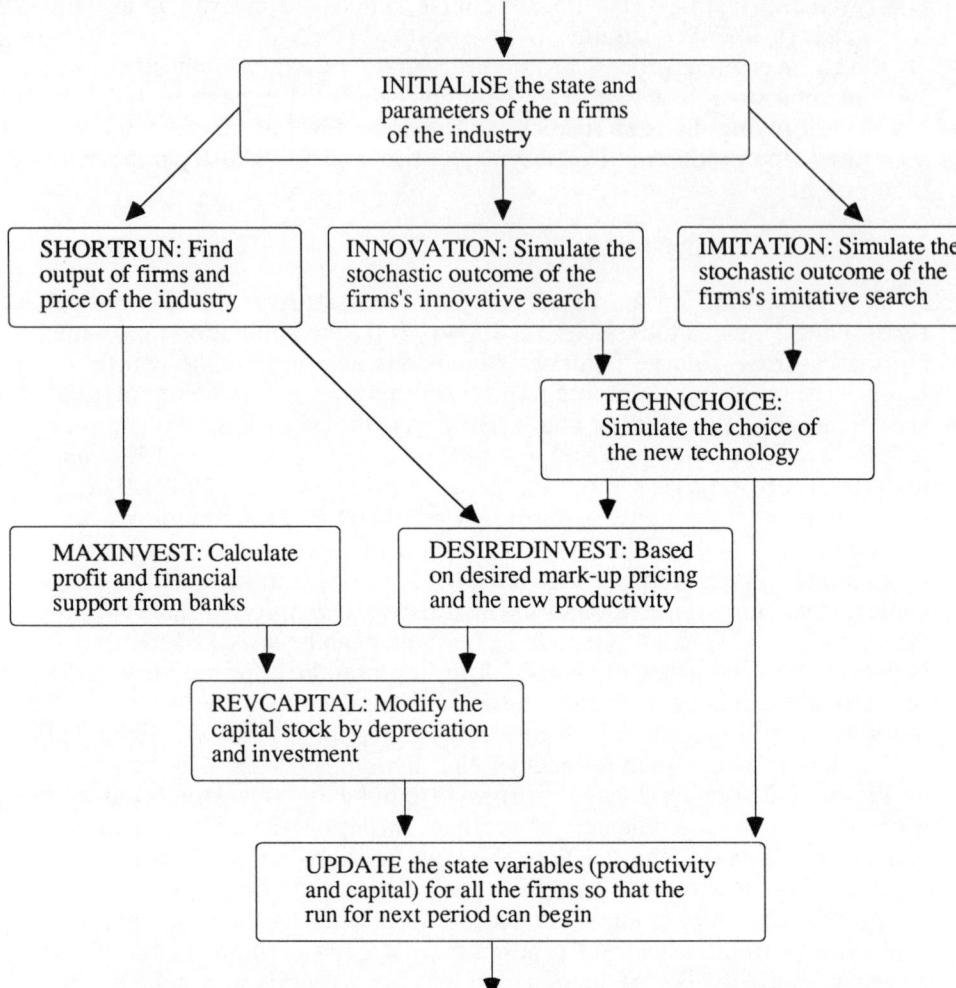

Figure 4.2. The basic procedures of typical Nelson-and-Winter models

Figures 4.1 and 4.2 can serve as direct starting points for further discussions and for the programming exercises which are reported in the appendix to this book. Such an immediate jump to the details of programming is, however, not advisable. It can easily lead to idiosyncratic specifications of firm and market behaviour which means that the simulation results are of little or no value. It may also serve to block the possibility of exploring the mathematical properties of the computational system. For such reasons there is a need to make a more formal specification of the model.

4.3. A formal version of the computational structure of typical Nelson-and-Winter models

Before the Nelson-and-Winter simulation models were designed, Winter (1964; 1971/93) was engaged in theoretical modelling and this work supplies much of what is needed for a formal specification of the later programming exercises. After much of the common work was performed, Winter (1984/91; 1986) also made presentations of the Nelson-and-Winter models which provide the best available specifications (but we also find specifications in, e.g., Nelson and Winter, 1978; 1982).

However, even if this material is available, there is a need to rethink the computational structure of typical Nelson-and-Winter models for two reasons. First, such a specification may serve as a common frame of reference for evolutionary-economic research. Such a reference is very important as a means of providing concerted action rather than Babylonic confusion.[25] Second, the algorithmic orientation of the specification may serve to ease the development of Artificial Economic Evolution. Third, formalisation of a computational process may serve to create an interface between researchers working with on the one hand formal-theoretical analysis and on the other hand numerical computer simulation.

Even if the rough description of the basic structure of the most important type of Nelson-and-Winter simulation models can be expressed by mathematical equations, I prefer to apply an algorithmic notation (see A.1.1) which emphasises the relationship between formal specifications and programming language specifications (some of which are listed in A.2 and A.3). Three major characteristics should be mentioned. First, the specification is made in the form of a series 'assignment statements' (':=' or '←') which denote change of the computer variables in successive order (from 4.1 to 4.22).[26] The creation of data structures and the modification of state variables are sometimes specified as 'definition statements' ('≡'). Second, the specifications are made in terms of a language which is able to formulate *all* possible computations (according to the Church-Turing thesis). It uses sequences of statements (do A; do B;...), iteration (while < condition > do A;) and selection (if < condition > then A;). The power of the notation will not be seen much in most of the following specifications since it concentrates on fairly simple issues. However, the computer procedures in the appendix use practically the same notation, and here it is clearly needed even if only simple procedures have been recorded. Third, variables are named with time subscripts. This does not mean that the computer (or the firm) allocates new memory space for all these variables. Instead a new value is assigned to an old variable. The subscripts make it easier to distinguish between different types of variables (scalars and vectors), and they are necessary for mathematical treatment.

In making the formal description of the simple Nelson-and-Winter model, it has been natural to spell out several details which are implicit in

Nelson and Winter's accounts (and especially their poorly documented book).[27] However, it must be emphasised that all descriptions operate at a certain level of abstraction which leaves many kinds of detail open until the phase of implementation (appendix). In the present case many important decisions about the specification of central functions of the Nelson-and-Winter model are left out (except for occasional remarks and footnotes). Of special importance are the problems relating to the stochastic functions describing the processes of innovation[28] and imitation. Actually, it is these functions which to a large extent determine the outcome of the subsequent simulations. In any full treatment of the Nelson-and-Winter models these functions should be spelled out and discussed. The goal of the present exercise is, however, to prepare the ground for a discussion of Schumpeter's work in a modern evolutionary-economic context. This purpose suggests a condensed exposition which, e.g., has to leave out the controversy with textbook neoclassicism which occupies too much of the space in Nelson and Winter's writings.[29]

4.3.1. Inheritance from period $t - 1$

We have an industry composed of firms producing a homogeneous product by means of homogeneous physical capital where the productivity is determined by techniques. To describe the state of the computation between the periods we have

$$K_t \equiv (K_{1t}, \ldots, K_{it}, \ldots, K_{nt});$$

$$A_t \equiv (A_{1t}, \ldots, A_{it}, \ldots, A_{nt});$$

where K_t and A_t are arrays (vectors) describing the state of the system in the beginning of period t; K_{it} is capital stock of firm i at time t; n is the fixed number of firms in the industry; A_{it} is productivity of capital (the output/capital ratio).

4.3.2. Short-run behaviour

Given the state of a firm, we compute its output which is simply determined by its capital productivity and its capital stock (4.1); at the same time we have the total production of the industry (4.2). All output is sold but its price is determined by the (simplistic) demand conditions which take price to be determined by the output of the industry in the period (4.3). We may now find the sales of each firm (4.4) as well as production costs (4.5) and the unconventional variables: innovative and imitative search costs ((4.6), (4.7)). Finally, we calculate the net return (4.8). In formal terms this series of computations can be described as:

$$Q_{it} := A_{it} K_{it}; \tag{4.1}$$

$$Q_t := \sum_i Q_{it}; \tag{4.2}$$

$$P_t := D(Q_t); \tag{4.3}$$

$$S_{it} := P_t Q_{it}; \tag{4.4}$$

$$C_{it} := cK_{it}; \tag{4.5}$$

$$R_{it}^n := r^n K_{it}; \tag{4.6}$$

$$R_{it}^m := r^m K_{it}; \tag{4.7}$$

$$Z_{it} := S_{it} - \left(C_{it} + R_{it}^m + R_{it}^n\right); \tag{4.8}$$

where Q_{it} is quantity of output of firm i at time t; Q_t is aggregate output of the industry; P_t is price per unit of output; $D(\ )$ is the demand-price function for the output; S_{it} is total sales; c is production cost per unit of capital (including capital rental);[30] C_{it} is total production cost for firm i in period t; r^n is cost of innovative activities per unit of capital; R_{it}^n is total innovation cost; r^m is cost of imitative activities per unit of capital; R_{it}^m is total imitation cost; Z_{it} is net return or economic profit.

4.3.3. Productivity change through innovation and imitation

The computations of productivity change are central to the results of the Nelson-and-Winter simulations of evolutionary processes. However, this part of the model is relatively poorly documented by Nelson and Winter. It includes two broadly similar components. First, we have innovation: the innovative search effort determines the probability of finding an innovation (4.9), while a complex probability function[31] determines whether search will be successful in period t (4.10). The concrete outcome of the successful search is determined by another sampling from a distribution of technological opportunities (4.11). Second, we have imitative search whose success or failure are determined in a similar way as in the case of innovation ((4.12), (4.13)). However, the outcome of successful imitation is much simpler: it implies an access to the best-practice technique of the industry (4.14). Finally, the firm chooses between the existing technique and the new possibilities (4.15). Formally:

$$\gamma_{it}^n := d^n R_{it}^n; \tag{4.9}$$

$$\theta_{it}^n := \theta^n\left(\gamma_{it}^n\right); \tag{4.10}$$

$$\left.\begin{array}{l}\underline{\text{if}}\ \theta_{it}^{n} = \underline{\text{true}} \\ \underline{\text{then}}\ A_{it}^{n} := \psi\left(A^{\text{init}}, \varphi, t, A_{it}\right) \\ \underline{\text{else}}\ A_{it}^{n} := A_{it};\end{array}\right\}^{32} \quad (4.11)$$

$$\gamma_{it}^{m} := d^{m} R_{it}^{m}; \quad (4.12)$$

$$\theta_{it}^{m} := \theta^{m}\left(\gamma_{it}^{m}\right); \quad (4.13)$$

$$\left.\begin{array}{l}\underline{\text{if}}\ \theta_{it}^{m} = \underline{\text{true}} \\ \underline{\text{then}}\ A_{it}^{m} := \max[A_{1t}, \ldots, A_{it}, \ldots, A_{nt}] \\ \underline{\text{else}}\ A_{it}^{m} := A_{it};\end{array}\right\} \quad (4.14)$$

$$A_{i,t+1} := \max[A_{it}, A_{it}^{m}, A_{it}^{n}]; \quad (4.15)$$

where γ_{it}^{n} is the probability of getting a 'draw' in the lottery of innovation which depends on the similar probability per unit of search costs, d^{n}; θ_{it}^{n} is a variable which is <u>true</u> in case of an innovative 'draw' and otherwise <u>false</u>; $\theta^{n}(\ )$ is the probabilistic function which determines whether a 'draw' will occur in period t; A_{it}^{n} is the outcome of innovative activities; $\psi(\ )$ is the probabilistic function which determines the outcome of an innovative 'draw'; A^{init} is the initial value and φ is an expression of the time-dependence of the mean of the log-normally distributed A_{it}^{n}; γ_{it}^{m} d^{m} and A_{it}^{m} are the expressions for imitative activities which correspond to the similar expressions for the process of innovation; max[ ] is the function choosing the best among the available production techniques.

4.3.4. Change in physical capital[33]

The sequence of computations about the change in physical capital is relatively straightforward. The financial constraint (4.17) is calculated on the basis of the profitability of the firm (4.16) as well as bankers' financial rules. The desired investment (4.20) depends on the market share of the firm (4.18), the mark-up factor (4.19) and the depreciation rate. Actual investment implies a constrained fulfilment of desires (4.21) and the new capital stock is found by taking into account the old capital stock and depreciation (4.22). Formally:

$$\pi_{it} := \frac{Z_{it}}{K_{it}}; \qquad (4.16)$$

$$I_{it}^{\max} := G(\pi_{it}, b) \cdot K_{it};^{34} \qquad (4.17)$$

$$\mu_{it} := \frac{Q_{it}}{Q_t}; \qquad (4.18)$$

$$\rho_{it} := \frac{P_t}{c/A_{i,t+1}}; \qquad (4.19)$$

$$I_{it}^{\text{des}} := H(\rho_{it}, \mu_{it}, \delta) \cdot K_{it}; \qquad (4.20)$$

$$I_{it} := \max\left[0, \min\left[I_{it}^{\text{des}}, I_{it}^{\max}\right]\right]; \qquad (4.21)$$

$$K_{i,t+1} := I_{it} + (1-\delta)K_{it};^{35} \qquad (4.22)$$

where π_{it} is economic profit per unit of capital; $I_{it}^{\max}$ is maximum gross investment; $G(\ )$ is the function which determines maximum gross investment per unit of capital; b is the ratio of external financing to economic profit; δ is the physical depreciation rate per unit of capital; μ_{it} is the market share; ρ_{it} is the price/unitcost ratio; I_{it}^{des} is desired gross investment; $H(\ )$ is the function which determines desired gross investment per unit of capital; I_{it} is gross investment; max[ ] and min[ ] are the well-known functions which take the maximal and the minimal of their arguments.

4.3.5. State of the system at the beginning of period $t + 1$

At the end of the computations performed for period t we have the updated state of the system organised as

$$\boldsymbol{K}_{t+1} \equiv (K_{1,t+1}, \ldots, K_{i,t+1}, \ldots, K_{n,t+1});$$

$$\boldsymbol{A}_{t+1} \equiv (A_{1,t+1}, \ldots, A_{i,t+1}, \ldots, A_{n,t+1});$$

To prepare the system for the next round of computations we change the arrays of state variables as well as the time counter:

$$\boldsymbol{K}_t := \boldsymbol{K}_{t+1};$$

$$\boldsymbol{A}_t := \boldsymbol{A}_{t+1};$$

$$t := t + 1;$$

We may now go to section 4.3.2 and perform computations (4.1)–(4.22) once more.

4.3.6. Implementing the formal specifications

The thesis underlying the present report is that programming languages and concrete programs are important for economic-evolutionary analysis for two reasons. First and most obviously, the programming approach allows the researcher to create artificial worlds in which he can launch and study well-defined but highly complex 'evolutionary' processes. In some cases, the researcher may gain his first rough concept of evolution through such an interaction with a computer system. Second, the researcher may use the notation of a programming language as well as the results of computer simulations as a means of communication with other researchers. The programming notation may give clarity and power to the communication and in the future this may be of major importance for the development of evolutionary analysis.

The sceptical economist may ask whether attempts to clarify and imitate evolutionary processes by means of computer simulation really make any difference. What can we possibly learn from simulation that we did not know already? This question is based on the twin propositions that a computer does what it is programmed to do and that a simulation cannot be better than its underlying assumptions ('garbage in, garbage out'). The apparently obvious conclusion is that we cannot learn anything: computers and simulations are just means of expressing given ideas (like the above specifications).

While the propositions are correct, they do not lead to the conclusion that computer-implementable models are largely irrelevant. First, we have the well-known fact that it is often quite difficult to discover the consequences of a correct set of propositions; here simulation will often help us. Second, and more importantly, we have the effect of trying to express poorly understood processes in a form which is implementable on a digital computer. In cognitive science we see how the long-term stalemate in the verbal modelling of cognitive phenomena has been overcome and is followed by a creative (and partly confusing) scientific boom which is directly and indirectly related to Artificial Intelligence.[36] Here we see the crucial importance of simple standard cases (like chess-playing and theorem proving), core algorithms and data structures (lists, production systems, heuristic search, learning, etc.), and specialised programming languages (LISP, PROLOG, etc.).[37] Similarly, we may expect that Artificial Evolution will create its standard cases, algorithms, data structures and programming languages.[38]

Against this background and when we consider how to program Nelson-and-Winter models, we may say that the above specification is very incomplete. After such a general specification, the next task is to program the data structures and implement the procedures (appendix).

Here a large number of new problems of major importance for the evolutionary process emerge. Finally, we come to the concrete 'running' of the procedures. In this step we specify the maximum number of firms and periods to be studied. We also call the initialisation procedure which defines the initial values of parameters and some variables. At this point it is, e.g., possible to specify

1. Whether a standard search space should be used or whether a new search space with a certain number of points (and, perhaps, a certain topography) should be created.
2. Whether standard values for parameters and variables should be used or whether special values are to be applied.
3. Whether the special initialisation system defined by Nelson and Winter (1982, ch. 13) should be used; here four different parameters may be set low or high by means of binary code ('NW13 = [1,0,1,1]').
4. How to cope with 'random' events. Here we normally apply a (pseudo-) random-number generator, which is central to the behaviour of the model run. If the 'seed' of the sequence of random numbers is not changed, then the result for a given number of firms and periods will always be the same.

As a starting point such questions have been answered in a way which is close to the actions and computations defined by Nelson and Winter (1982, ch. 12). In other words, the present implementation is designed as a compromise between

(1) closeness to the Nelson and Winter formulations,
(2) ease of data access and revision of parameters and procedures,
(3) access to powerful mathematical tools and programming languages.

Thus, the implementation is mainly designed as a simple base for further work rather than as an embodiment of novel viewpoints. To the extent that the system functions and is repeatedly used, it will give novel results through a process of trial and error (see A.1.2). We have, however, to start by simple exercises which demonstrate the basic parameters and the basic behaviour of typical Nelson-and-Winter models. Figure 4.3 gives an example of the initial exercises, namely the development of the productivity levels of 4 firms for 16 periods under a given distribution of innovative and imitative chances (and a given setting of the random-number generator: '_seed := 2'). Initially all firms have equal market shares, equal research effort and a productivity level of 0.16.[39] However, already in the second period one firm has luck in its innovative R&D and is able to increase its productivity to 0.21. This means that it wants to expand its capital stock. In the fourth period the situation is changed since another firm makes an innovation. And later both innovations and imitations take place.

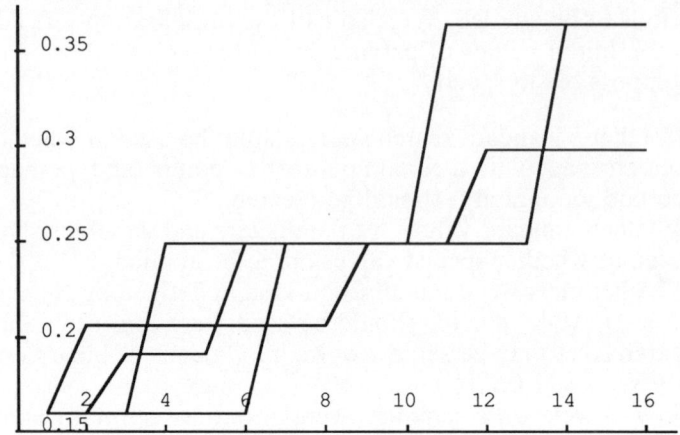

Figure 4.3. Productivity levels of 4 firms for 16 periods under a given distribution of innovative and imitative chances (_seed := 2). (Original is in colour to distinguish between firms)

Figure 4.3 demonstrates that the overall productivity level of the industry develops as a result of innovation and imitation. In the latter case a weak firm may suddenly get access to the productivity frontier. Whether this is the case depends on the specification of the possibilities of imitation just as the innovative possibilities are specified by the search space as well as the probability of obtaining an innovative 'draw'.

The market shares of firms and the concentration ratio of the industry depend on the relative weight of the two forces. On the one hand a lucky firm may obtain a competitive advantage through innovation. If the other firms are able neither to innovate nor to imitate, then this firm may obtain a monopoly position. This tendency is, however, weakened by the monopolistic practices of dominant firms with little incentive to invest (section A.2.7). On the other hand there is a (decreasing) chance that a weak firm overcomes its weakness through imitation and, to a smaller extent, through innovation.

The development of market shares in a specific run is depicted by figure 4.4. It demonstrates the volatility of the market positions under the chosen conditions. A lucky innovator who alone obtains a high productivity level will increase his capacity and production as a result of the increased productivity as well as of his increased investment activities.

Exploring the process of 'Schumpeterian competition' 117

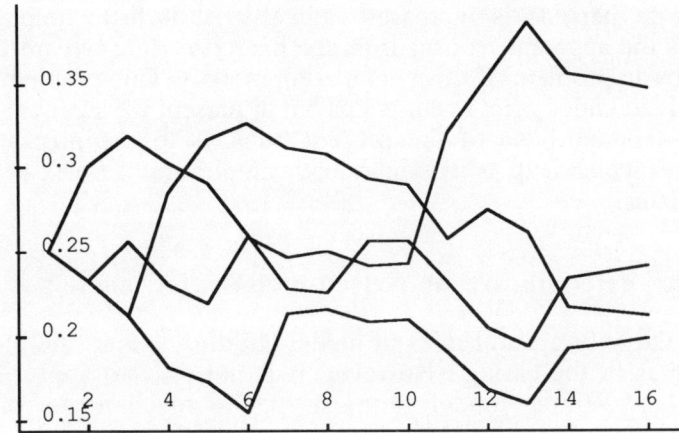

Figure 4.4. Market shares of 4 firms for 16 periods
under another distribution of innovative and imitative chances (_seed := 2).
(Original is in colour to distinguish between firms)

Figure 4.4. suggests that no stable pattern is available. But in figure 4.5. another series of random events (based on _seed := 1) appears to give another result with an early winner and three unlucky firms among which one escapes the downward trend after period 10.

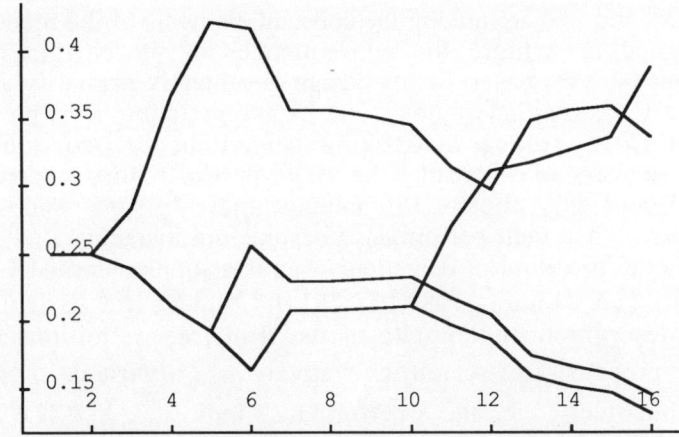

Figure 4.5. Market shares of 4 firms for 16 periods
under a certain distribution of innovative and imitative chances (_seed := 1).
(Original is in colour to distinguish between firms)

This limited experiment serves to demonstrate that the exogenous random-number generator may influence the behaviour of Nelson-and-Winter models heavily while the endogenously generated behavioural patterns may be hidden by this 'noise'. To study the behaviour as the number of periods is increased radically is of little help since this removes the apparent realism from the exercise. Here we are confronted with a basic problem of most simulation models. This problem has often been solved under strict assumptions but at present we shall not explore it further. The purpose of the present book is to explore models and languages which help us to express both simple and complex evolutionary mechanisms.

4.4. The extendibility of Nelson-and-Winter models

One of the only general rules of model-building is that 'the proof of the pudding is in the eating'. However, it is not easy to apply the simple Nelson-and-Winter model to most of the mechanisms one would intuitively connect to evolutionary economics. Therefore, the use of the model is largely a question about its extendibility. Before we turn to an evaluation of the model, it is thus important to consider some extensions.[40]

4.4.1. Satisficing behaviour[41]

In contrast to other of Nelson and Winter's models the type presented in sections 4.2.2 and 4.3 does not include satisficing behaviour in an explicit way. This is clearly demonstrated by the fact that the propensities to search (r^n and r^m) are among the constant elements of the model. Thus it is designed to explore the consequences of more rigid modes of behaviour but the reason for its design is probably primarily an attempt to simplify the simulation model and its interpretation. It is, however, not difficult to reintroduce satisficing behaviour, i.e., to operate with decision-makers who consider the major part of reality as consisting of 'givens', and only attempt to challenge these 'givens' and search for alternatives when their performance become unsatisfactory.

To model this kind of behaviour, we, as a supplement to K_{it} and A_{it}, introduce (see Winter, 1984/91, 281 f.) three extra variables in the simple description of the state of the firm (see definitions below): a variable propensity to search innovatively, r_{it}^n, a variable propensity to search imitatively, r_{it}^m, and a performance indicator, X_{it}. The latter (see 4.23) is defined as a distributed lag function of firm i's profit rate, and when it is compared with the industry's average profit rate, we have a measure of the aspiration level of satisficing behaviour (used in (4.24) and (4.25)).[42] If this level is obtained, no change is made in innovative and imitative search policies. If not, a step towards the industry average is made. Furthermore, the firm decides to commit itself to the new policy for some time in order to allow the policy to 'work itself out'. This decision is reflected in the increase of X_{it} with some amount, Δ, when a change in search strategy has been made:[43]

$$r_t^n \equiv \left(r_{1t}^n, \ldots, r_{it}^n, \ldots, r_{nt}^n\right);$$

$$r_t^m \equiv \left(r_{1t}^m, \ldots, r_{it}^m, \ldots, r_{nt}^m\right);$$

$$X_t \equiv \left(X_{1t}, \ldots, X_{it}, \ldots, X_{nt}\right);$$

$$X_{it} := \phi X_{i,t-1} + (1-\phi)\pi_{it}; \tag{4.23}$$

$$\left.\begin{array}{l} \text{if } X_{it} \geq \overline{\pi}_t \\ \quad \underline{\text{then}} \ r_{i,t+1}^n := r_{it}^n \\ \quad \underline{\text{else}} \ r_{i,t+1}^n := \beta r_{it}^n + (1-\beta)\overline{r}_t^n + u_{it}^n; \\ \quad X_{it} := X_{it} + \Delta; \end{array}\right\} \tag{4.24}$$

$$\left.\begin{array}{l} \text{if } X_{it} \geq \overline{\pi}_t \\ \quad \underline{\text{then}} \ r_{i,t+1}^m := r_{it}^m \\ \quad \underline{\text{else}} \ r_{i,t+1}^m := \beta r_{it}^m + (1-\beta)\overline{r}_t^m + u_{it}^m; \\ \quad X_{it} := X_{it} + \Delta; \end{array}\right\} \tag{4.25}$$

where X_{it} is the perceived performance of firm i; ϕ is a parameter which determines the inertia of the performance indicator $(0 < \phi < 1)$; $\overline{\pi}_t$ is current industry capital-weighted average profit; $\overline{r}_t^n$ and $\overline{r}_t^m$ are the capital-weighted industry average search efforts (policies) of the period; β is a parameter which determines the inertia of the firm's innovative and imitative search policies; u_{it}^n and u_{it}^m are random variables[44]; Δ is a fixed increment in the perceived performance introduced to stabilise a new search policy.[45]

4.4.2. Exit and entry

There are good reasons to believe than an important part of evolutionary processes are related to the liquidation of some firms and the emergence of others. The economist who has put most emphasis on this aspect of the evolutionary process is the verbal Mark I model of the young Schumpeter while the old Schumpeter emphasises the more routine-like technical change performed by large corporations in his verbal Mark II model.[46] The evolutionary simulations which can be created on the basis of the standard Nelson-and-Winter model ((4.1)–(4.22)) and of its extension with satisficing behaviour ((4.23)–(4.25)) are clearly more related to the Schumpeter Mark II model than to the Schumpeter Mark I model. This bias is immediately discernible from the fact that in these Nelson-and-

Winter models the industry's number of firms, n, is a constant. There are, however, some possibilities for introducing exit and entry to as well as industry creation and thus to operate with a variable number of firms, n_t. Such changes will not necessarily change the basic flavour of the Nelson-and-Winter models, but they open the way for a discussion about the differences between Schumpeter Mark I and II. The extension of the Nelson-and-Winter model in this direction is most clearly stated by Winter (1984/91, 283-288) but the elements were also present in earlier versions of Nelson-and-Winter models.

The story of firm exit from the industry is relatively simple. Here we have two cases. First, if a firm invests less than the physical depreciation of the capital stock over several periods, it will in the end fail with respect to a minimum capital stock level, K^{min} (which reflects the existence of sharply increasing returns to scale up to this threshold). Second, the perceived performance of the firm, X_{it} (see (4.23)), may fall below a critical negative level, X^{min}. This means that the firm relatively persistently is showing an unacceptable performance. In both cases the firm is liquidated and its capital is scrapped:

$$\left. \begin{array}{l} \underline{\text{if}} \left(K_{i,t+1} < K^{min} \underline{\text{ or }} X_{i,t} < X^{min} \right) \\ \underline{\text{then }} K_{i,t+1} := 0; \end{array} \right\}_{47} \qquad (4.26)$$

Firm entry to the industry raises some more complicated problems which relate to the profit-seeking activities of actors which are not producing the product of the industry we are modelling. In a highly stylised manner, we may articulate these activities in terms of costly innovative and imitative search like the activities we find within the industry ((4.7) and (4.8)[48]). Taken as an aggregate, external actors are involved in search activities (R&D) which will occasionally give results with respect to the production technique of the industry under study. This leads to some number ($N \geq 0$) of external actors who in each period have success in achieving relevant information about the techniques of the industry. We may assume that this number have a Poisson probability distribution with a parameter proportional to the cost of industry-external search.

Whether a 'draw' by an external actor leads to entry to the industry depends on whether the successful searcher thinks that the obtainable productivity allows a profit rate which exceeds some critical level π^{emin} (the 'entry barrier' level). The relevant information is not fully available and therefore a random error in the firm's evaluation of the production conditions of the industry is introduced. If a firm decides to enter, then it applies the technique it has found and establishes an initial capital stock, initial search policies and an initial level of performance. When we take into account the difference between innovative and imitative search of the external actors, we obtain the following statements:[49]

$$\boldsymbol{A}_t(t) \equiv (A_{1t},\ldots,A_{it},\ldots,A_{n(t),t});$$

$$\boldsymbol{K}_t(t) \equiv (K_{1t},\ldots,K_{it},\ldots,K_{n(t),t});$$

$$\boldsymbol{r}_t^n(t) \equiv \left(r_{1t}^n,\ldots,r_{it}^n,\ldots,r_{n(t),t}^n\right);$$

$$\boldsymbol{r}_t^m(t) \equiv \left(r_{1t}^m,\ldots,r_{it}^m,\ldots,r_{n(t),t}^m\right);$$

$$\boldsymbol{X}_t(t) \equiv \left(X_{1t},\ldots,X_{it},\ldots,X_{n(t),t}\right);$$

$$\gamma_{it}^{ne} := d^{ne} \cdot R^{ne}; \qquad (4.27)$$

$$N_t^{ne} := \theta^{ne}\left(\gamma_{it}^{ne}\right); \qquad (4.28)$$

repeat N_t^{ne} times

$$\left.\begin{aligned}
&A_e := \psi^e\left(A^{init},\varphi,t\right); \\
&\text{if } \left(P_t A_e - c > \pi^{emin} + u_{et}\right) \text{ then} \\
&\quad A_{n+1,t+1} := A_e; \\
&\quad K_{n+1,t+1} := \xi\left(K^e,\sigma^k,K^{min}\right); \\
&\quad r_{n+1,t+1}^n := \bar{r}_t^n + u_{it}^n; \\
&\quad r_{n+1,t+1}^m := \bar{r}_t^m + u_{it}^m; \\
&\quad X_{n+1,t} := \Delta; \\
&\quad n_t^{id} := n_t^{id} + 1;
\end{aligned}\right\} \qquad (4.29)$$

$$\gamma_{it}^{me} := d^{me} \cdot R^{me}; \qquad (4.30)$$

$$N_t^{me} := \theta^{me}\left(\gamma_{it}^{me}\right); \qquad (4.31)$$

repeat N_t^{me} times

$$\left.\begin{array}{l}A_e := \max[A_{1t},\ldots,A_{it},\ldots,A_{nt}]; \\ \underline{\text{if }} \left(P_t A_e - c > \pi^{emin} + u_{et}\right) \underline{\text{ then}} \\ \quad A_{n+1,t+1} := A_e; \\ \quad K_{n+1,t+1} := \xi\left(K^e, \sigma^k, K^{min}\right); \\ \quad r_{n+1,t+1}^n := \bar{r}_t^n + u_{it}^n; \\ \quad r_{n+1,t+1}^m := \bar{r}_t^m + u_{it}^m; \\ \quad X_{n+1,t} := \Delta; \\ \quad n_t^{id} := n_t^{id} + 1;\end{array}\right\}50 \qquad (4.32)$$

where the following new notation is introduced: $K_t(t), A_t(t), r_t(t), r_t(t), X_t(t)$ are the arrays describing the state variables but their number of elements $(n(t))$ have now become time-dependent; each time a new firm is created the number of elements in all the arrays are increased by one; N_t^{ne} and N_t^{me} are the numbers (each ≥ 0) of potential entrants at the end of period t due to industry-external innovative and imitative search; $\theta^{ne}(\ )$ and $\theta^{me}(\ )$ are the functions which determine the number of entrants for a period and ensure that this number has a Poisson probability distribution; the means of the distributions are γ_{it}^{ne} and γ_{it}^{me}; they are determined by the (for simplicity unchanging) levels of industry-external search expenditure which may lead to information about applicable production techniques, R^{ne} and R^{me}, as well as be the efficiency coefficients of these efforts, d^{ne} and d^{me}, i.e., the probabilities that a unit of innovative and imitative search expenditure will lead to success; A_e is a local variable denoting the capital productivity which is obtainable through the technique found by a potential entrant, e; $\psi^e(\ )$ is the probabilistic function determining the outcome of an innovative 'draw' by an external actor (see (4.11)); π^{emin} is the minimum level of profit per unit of capital required by the potential entrants (and their banks); u_{et} is a random variable[51] which represents an error term in the information and calculation of a potential entrant; $A_{n+1,t+1}$, $K_{n+1,t+1}$, $r_{n+1,t+1}^n$, $r_{n+1,t+1}^m$, $X_{n+1,t}$ are newly created state variables for the new firm of the industry; $\xi(\ )$ is the function which probabilistically determines the initial capital stock of the entrant with a mean value K^e, a standard deviation σ^k, and a downward truncation determined by K^{min}; n_t^{id} is a variable which is used for the naming (by an identification number) of new firms of the industry.

4.4.3. Industry creation

When the problems of entry have been dealt with, there is no large step to start a simplified study of industry creation. Basically, the simulation model ((4.1)–(4.32)) can be used although we have to introduce minor modifications. The study of industry creation is thus basically a case of running the model after a little reprogramming and the definition of special parameter values. In the present context it is, however, more helpful to demonstrate how the above exposition of entry to the industry illustrates a procedure which may be extended to a treatment of industry creation.

In principle we just start with $n_t^{id} = 0$ and then we let the search activities of the external actors find the industry and start it up on the basis of (4.27) to (4.29).[52] In the beginning one or more founders or *Gründer* are successful in the background innovative search. In other words, he finds a production technique allowing him to produce the product. To decide whether to invest and start producing the product he needs some initial information about the price of the product as well as of production inputs. A justification of the application of the entry model may be found in the fact that the demand-price of a new product is often judged on the basis of an existing product which is functionally more or less equivalent to it. Input prices may also be assumed to be known. Thus, the founding of an industry is dependent upon 1) the level of background search, 2) the demand-price for the first products delivered, 3) the production technique found and thus the initial profitability. Further factors are: 4) the minimum scale of investment and production and 5) the financial conditions of potential entrants, including bankers' rules about finance of industry creation (and thus their way of coping with asymmetric information between lender and borrower in the case of founders).

We now turn to the evolution of the newly created industry.[53] In the industry there is initially a large profit rate (at least enough to overcome the entry barrier, see the condition in (4.29)). The rate of expansion of the firms of the industry as well as the rate of entry to the industry is high (relative to later). The initial search efforts of the firms of the industry are low according to the rules of satisficing behaviour ((4.24)–(4.25)). Consequently, the level of capital productivity is low. However, the expansion of the output of the industry will diminish profits and this will lead to negative evaluations of the performance of the firms. This in turn will lead to an increase in innovative and imitative search efforts (in steps because of the heuristic rule allowing some time for new search strategies to work themselves out). In any case, we see an increase in productivity which eases the pressure on profit.

With respect to the number of firms of the industry, n_t, we see an initial increase. However, since there is no feedback from the profitability of the industry to the search efforts of the external actors in the present version of the model, the growth in the number of firms may be relatively slow. In any case, by entry and internal expansion the industry will sooner or later reach a capacity level and a demand-price which means liquidation for some firms (4.26). Here the casualties are firms with below-average productivity. The survivors are winners in the sense that exit leads to a scrapping of part of the productive capacity of the industry and thus to an increase in profits.[54]

4.4.4. Product quality, product innovation and imitation

A central weakness of the Nelson-and-Winter models is that they deal solely with production techniques and thus delimit their focus to process innovation and imitation. In the actual evolutionary-economic processes, product innovation and imitation appears to be much more important, especially for the activities we call R&D. However, the Nelson-and-Winter scheme is flexible enough to allow a crude treatment even of these issues. One possibility, developed by Gerybadze (1982, chs. 3-5), is to expand the application of the scheme to the interaction of firms which sells and buy a certain type of product. Another possibility is to emphasise that while

> Winter argues persuasively for adaptive modeling of firm behaviour ... [t]he case for adaptive consumer modeling is stronger, since consumers are less able to bear the cost of computing optima and since inefficient consumers will not be driven out of existence. (Smallwood and Conlisk, 1979)

Even if the latter possibility represents a major break with the Nelson-and-Winter family of models, it is enlightening to make a short excursion into the area of stochastic quality attributes.

An excursion on consumers and product quality

To start with, brand quality can be considered as the probability that a single specimen of the brand received by the buyer will not reveal itself as being of an unsatisfactory quality during the normal period of use. High quality means a low probability for a unit of the brand to show a 'breakdown', an inadequate performance seen from the viewpoint of the buyer (Smallwood and Conlisk, 1979, 3). The task of the buyer is to learn about the breakdown probability for each brand and choose accordingly.

In the model, buyers tend to follow simple adaptive strategies with respect to such goods. In this way they try to cope with the fact that (1) they are poorly informed in comparison to sellers, (2) any attempt to overcome the asymmetric information would run into their competence-difficulty gap (Heiner, 1988) which is especially marked with respect to issues which involve stochastic product quality, (3) a relatively consistent following of simple heuristic rules may accumulate knowledge of the credence qualities of brands—either in the individual or the buyers as a group.

The most obvious strategy is that of satisficing. which we have until now seen from the viewpoint of firms, but it seems even more obvious in the modelling of the behaviour of non-specialist buyers. For them satisficing behaviour is based on two basic rules which are applied by each buyer.

> Rule I: if the brand chosen by the buyer in the last purchase has not shown unsatisfactory behaviour to him, then the same brand is chosen in the present purchase.
> Rule II: if the brand chosen by the buyer in the last purchase has shown unsatisfactory behaviour, then a search is performed in order to find a better brand to be chosen in the present purchase.

To the extent that rule II is specified in a way which takes into account other buyers' experiences with the different brands, then a collective learning process about brand quality may take place. The learning among buyers may, however, not be sufficient if a 'breakdown' is a rare but very grave event. In this case we may have to introduce another rule of social learning:

> Rule III: if the brand chosen by one of the other buyers in the last purchase has shown unsatisfactory behaviour and if this becomes known to the buyer, then a search is performed in order to find a better brand to be chosen in the present purchase.

To the extent that this rule is not sufficient we come to

> Rule IV: if a basic doubt exists about a grave form of breakdown and if this doubt cannot be removed by the seller, then a third party (e.g. the government) is introduced in order to ensure that all units of all brands have a minimum level of quality.

The rules of satisficing behaviour presuppose to a larger or smaller degree that the credence qualities emphasised by different buyers are (to some extent) the same. Rules III-IV raise still stronger demands on buyers' ability to define and communicate about credence qualities. Furthermore, we see that the rules presuppose a certain structure with respect to supply-side organisation. Underlying rules I-III is the assumption that the brands of the different sellers show a well-defined and stable (or only slowly changing) probability of unsatisfactory performance.

The behaviour of a simplified system based on rules I-II can be described in terms of a Markov process. The core idea is that buyers' experiences with the units of the brands in their possession during a particular period of time determine the market shares of the different brands in the next period. This idea has been explored by Smallwood and Conlisk (1979).

Their standard model is based on the assumption that all brands are almost identical. They are identical with respect to price and any directly observable quality. But they differ with respect to credence quality. Each product lasts one period but during this period the probability of breakdown is dependent on the brand that the unit belongs to. Each of the n brands is characterised by a positive probability of breakdown. The probability that brand i used during period t will breakdown is b_{it}. For instance, this probability may be 0.2, meaning that from a large sample of units of brand i 80% will be satisfactory while 20% will show some sort of breakdown. b_t is the vector of breakdown probabilities and they may differ across brands. For simplicity, we shall normally assume that breakdown probabilities are non-changing through time.[55]

Each buyer buys one product in the beginning of each period, $t+1$. He cannot discern its quality by pre-purchase inspection but only by post-purchase experience of a breakdown. If no breakdown is experienced with the brand in his possession during period t, then he buys a unit of the same brand in the beginning of period $t+1$. If a breakdown is experienced with the brand in his possession during period t, then he performs a search for a better brand before he decides which brand to buy in the beginning of period $t+1$.

In the model the outcome of the search for a better-quality brand is stochastically dependent of the market shares of the different brands during period t. The market share of brand i in period t, m_{it}, is simply the fraction of buyers who have bought this brand at the beginning of the period. The situation for all brands is summarised by the market share vector, m_t. Assuming a decision-making process which to some extent reflects quality of brands, the market shares can be taken as a more or less precise indicator of credence qualities. If market shares are good indicators, the buyer would like to choose the brand with the highest market share. But in some cases there might be no correlation between market share and quality. In other cases the buyer would not know market shares but only the way they are reflected in the probabilities that he will meet other buyers who know the different brands or sellers who sell the different brands.

The dissatisfied buyer can have different degrees of confidence in the ability of the market selection process to detect high-quality brands and different abilities to recognise the market shares of different brands. The empirical result will be a stronger or weaker relationship between market shares and the choice of the dissatisfied buyer. Smallwood and Conlisk (1979, 5 ff.) rely heavily on this parameter, α, the sensitivity of brand choice to market shares. The dissatisfied buyer chooses the brand for the next period with probabilities that are proportional to the market shares. For instance, the probability that a dissatisfied buyer chooses brand j in period $t+1$ is proportional to $m_{jt}^\alpha \bigg/ \sum_{k=1}^{n} m_{kt}^\alpha$, for $\alpha \geq 0$. If $\alpha = 0$, the buyer has no confidence in market shares as an indicator of quality; instead he chooses a brand at random. If $\alpha = 1$, the buyer has an intermediate confidence in market shares as an indicator of quality; he chooses a brand with a probability which is equal to its market share. If $\alpha = \infty$, the buyer has maximum confidence in market shares as an indicator of quality; he simply chooses the brand with the highest market share.

We have now defined a Markov process which is only reflecting choices of the buyers. Such a Markov process will more or less rapidly converge towards a state of equilibrium where the market share of each brand has reached an equilibrium level where for brand i we have that $m_{i,t+1} = m_{it}$. The question is now how these equilibrium market shares reflect the credence qualities of the different brands of piglets. At the one extreme we may find that only brands of the highest quality have a non-zero market share. At the other extreme we may find that the market selection process has not been able to discern the high-quality brands from the low-quality brands. It is hardly surprising that the answer (which is given by the simple mathematics of Markov processes) is heavily dependent on the buyers' confidence in the quality signals of the market, α. The correspondence between quality and market shares peaks for $\alpha = 1$ and falls for higher and lower parameter values. Too much confidence in the market signals gives a large probability that brands which for random reasons started with a relatively high market share dominate the equilibrium state. Too little confidence in the market allows brands of different quality levels to coexist but with smaller market shares of weak-quality brands than in the initial situation.

The problematic character of the equilibrium state becomes even more obvious if we introduce more radical buyer reactions to quality breakdowns. We may, e.g., take into account that a buyer who has discovered a breakdown of a particular brand will not include this brand in his search for the next choice. This behaviour decreases the application of the best-performing brand which also shows breakdowns. The 'correct' behaviour would be not to overreact when facing a breakdown of a unit of a brand and give the same brand a chance in the next choice. The overreaction to breakdown can also occur if we allow longer memories about unsatisfactory performance and stronger communication between buyers.

Possible models of product innovation and imitation

We now return to the main line of argument: how can Nelson-and-Winter-like models be adapted to cope with product innovation and imitation. Here we are entering a zone which confronts us with a whole series of new questions and complexities which have not yet been solved. In the present context we have only the possibility of glancing at a few possibilities (some of which have not yet been explored in the literature).

We start by rethinking the change of production techniques which has hitherto been considered as the introduction of disembodied techniques, i.e., techniques which can be introduced without respect to the quality of the capital equipment. Now we introduce embodied technical change where the increased productivity can only be obtained by introducing capital equipment of a higher quality. We discuss embodied technical change in a two-industry framework (see figure 4.2) which provides one of the simplest ways of introducing product-variety and variety-creation in the Nelson-and-Winter scheme. Here we may think of the type of capital-using industry we have been studying in most of the present chapter as industry A. The industry which produces the capital equipment for industry A is called industry B. The possibilities to continue to increase productivity in industry A depend on the ability of the firms in industry B to continue to find product innovations and to expand production of them by accumulation in the innovating B-firm and imitation from other B-firms.

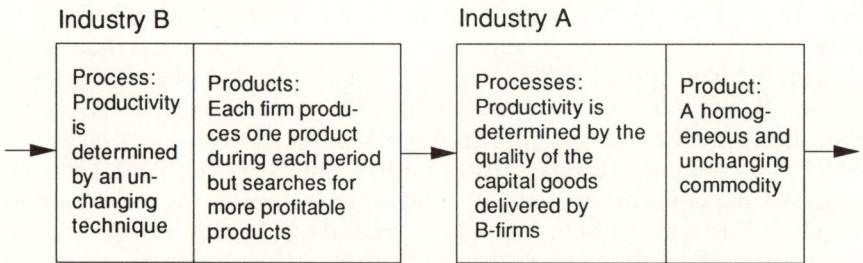

Figure 4.2. The set-up for product innovation and imitation in a capital equipment industry (industry B).

We start looking at this set-up from the viewpoint of the A-firms. In the previous models the capital equipment had no influence whatsoever on productivity because we assumed a disembodied form of technical change. Any new production technique could immediately be applied irrespective of the use of quite old vintages of capital. Now the A-firms can only introduce new techniques and thus higher levels of productivity in companionship with new capital equipment. Furthermore, we assume that this equipment is not produced by industry A but only by industry B. This removes the whole rationale for innovative and imitative search in industry A.[56] Instead the problem of the A-firms is to find the capital equipment with the best performance relative to its price.

Let us now approach the problem from the side of the B-firms. They have the possibility of producing a large but also largely unknown set of capital goods. To simplify, we assume that each B-firm uses its full production capacity to produce one type of capital good. Its production capacity is not dedicated to this type of good, and it may change to produce another good at the end of each period. All goods are produced with the same physical productivity. The only thing which makes them different is the way they function in industry A. The potential capital goods are assumed to be clearly distinct with respect to the productivity they ensure for the A-firms who are able to order them with respect to physical productivity per capital unit.

Given these assumptions, we may rethink the story told by statements (4.1)–(4.22). First of all, we need a description of the new situation:

1. We need to specify the set of B-firms. Furthermore, we must specify the set of possible products (capital goods) of industry B. We may, e.g., imagine that the search of the B-firms during a large range of periods may at the very best discover 100 different potential products each of which is provided with an identity number (see general definitions below).
2. To simplify further we assume that each existing B-firm will produce one and only one product during a period, t, and that the products produced in period $t + 1$ are decided by the B-firms during period t. This gives us a new element in our state description of industry B, namely a mapping between B-firms and products implemented as an array (see below).
3. Finally, we give characteristics to the potential products. For simplicity we assume that they are all identical with respect to their unit cost of production for the B-firms. However, the A-firms are able to discern between their quality because they lead to different output/capital ratios when applied in industry A. Let us distribute capital productivities to the 100 potential products, e.g., productivities distributed randomly between a very low and a very high level.

With these definitions we may formulate the process by which the capital goods produced and applied change over time.

1. To simplify we assume that the demand-prices of the A-firms for different capital goods strictly reflect their quality differences (4.33–4.34). This means that we assume full substitutability between the capital goods and full information on the part of the A-firms.[57]
2. Through their costs of innovative and imitative search (see (4.6)–(4.7), (4.9)–(4.10). (4.12)–(4.13)), the B-firms may succeed in having a 'draw' in the innovative and imitative 'lottery'. The probability of success depends on technological and institutional conditions which have to be specified in terms of their influence on the probability distributions.
3. The result of an innovative 'draw' (4.35) must be specified. We may, e.g., assume that the successful development of a new product is conditioned by the availability in the B-firm of 'tacit knowledge' which is related to earlier products produced by the firm. For this

reason, a firm is most likely to find products with a quality closely related to its earlier experience.
4. The result of an imitative 'draw' may be formulated in analogy to (4.14).
5. These specifications create a clear-cut decision-problem for the B-firms at the end of each period: to proceed with the good it is producing or to shift to another good. The possibility to make a change is the probabilistic outcome of search processes of the B-firms. They choose the possibility which has the highest quality in the eyes of the A-firms (4.37).

Through these modifications of the Nelson-and-Winter models we have core elements of an evolutionary model with product innovation and imitation. However, the missing elements are not simple to provide without greatly increasing the level of complexity (e.g., in terms of different vintages of capital in A-firms, scrapping rules of A-firms, etc.). Here we should just remark that some simple kinds of product innovation and imitation may be within the reach of a Nelson-and-Winter-type model. Some of the formal elements such a model should include are summarised in the following statements which mainly concern the firms in industry B:

$$f_B^{id} \equiv \{1,\ldots,i,\ldots,n\};$$

$$g^{id} \equiv \{1,\ldots,j,\ldots,m\};$$

$$G'_{Bit} \equiv (i, g_t) \underline{\text{where}} \ i \in f^{id}, g_t \in g^{id}; \ \text{or, simpler: } G_{Bit} \equiv g_t;$$

$$G_{Bt} \equiv (G_{B1t},\ldots,G_{Bit},\ldots,G_{Bnt});$$

$$A_A \equiv (A_{A1},\ldots,A_{Aj},\ldots,A_{Am}) \underline{\text{where}} \ j \in g^{id};$$

$$\underline{\text{constraint:}} \quad \underline{\text{if}} \ A_{Aj} > A_{Al} \ \underline{\text{then}} \ P_{jt} > P_{lt}; \tag{4.33}$$

$$P_{jt} := D_A(A_{Aj}, Q_{jt}); \tag{4.34}$$

$$\left.\begin{array}{l}\underline{\text{if}} \ \theta_{Bit}^n = \underline{\text{true}} \\ \quad \underline{\text{then}} \ G_{Bit}^n := \psi^g(\ldots) \\ \quad \underline{\text{else}} \ G_{Bit}^n := G_{Bit};\end{array}\right\}58 \tag{4.35}$$

$$\left.\begin{array}{l}\underline{\text{if }} \theta_{Bit}^{m} = \underline{\text{true}} \\ \underline{\text{then }} G_{Bit}^{m} := \max\left[A_{A}(G_{B1t}),...,A_{A}(G_{Bit}),...,A_{A}(G_{Bnt})\right] \\ \underline{\text{else }} G_{Bit}^{m} := G_{Bit};\end{array}\right\} \text{[59]} (4.36)$$

$$G_{Bi,t+1} := \max\left[A_{A}(G_{Bit}), A_{A}(G_{Bit}^{n}), A_{A}(G_{Bit}^{m})\right]; \qquad (4.37)$$

where f_{B}^{id} is a set of identifiers of B-firms;[60] g^{id} is a set of identifiers for potential capital goods/products; m is the number of potential products in all periods; G_{Bit} is the identification number of the capital good produced by firm i during period t; G_{Bt} is the array describing the state of the firms of industry B with respect to their product choices in period t; A_{A} is an array which to each potential capital good ascribes a capital productivity, seen from the users' (the A-firms) point of view; A_{Aj} is the productivity of the j'th capital good; j and l are two arbitrary goods; P_{jt} is the price of the j'th good; $D_{A}(\ )$ is the demand-price function for the inputs to industry A; θ_{Bit}^{n} and θ_{Bit}^{m} are described in (4.10) and (4.13); G_{Bit}^{n} and G_{Bit}^{m} are the products found by successful innovative and imitative search of a B-firm; $\psi^{g}(...)$ is the function which probabilistically determines the outcome of an innovative 'draw' of a B-firm.

The relevance of this formal apparatus is dependent on the large set of problems which can be explored with its help. However, such an exploration presupposes further developments of the Nelson-and-Winter framework which are beyond the limits of the present book. However, some applications in relation to non-linearities of the search space at the borders of national systems of innovation will be dealt with below.

4.5. Appreciating Nelson and Winter's modelling scheme

The foregoing presentation of a typical Nelson-and-Winter model as well as some extensions has been rather intricate but it has, hopefully, served its purpose: to demonstrate that Nelson and Winter have developed a computer-implementable concept of an evolutionary-economic process. This concept owes much to Simon and other contributors to Behavioural Economics but also includes several other ideas as well as a lot of synthetic work. Through the study of the extendibility of a typical Nelson-and-Winter model (section 4.4), it has also been demonstrated that it is not the specific assumptions underlying their standard model (sections 4.2.2 and 4.3) which should be used as the standard criteria to evaluate their research programme. The typical models as well as the concrete simulation experiments are (as claimed by Nelson and Winter) examples rather than primary results to be evaluated. Instead, we should try to evaluate Nelson and Winter's promotion and exploration of an extremely broad class of models which provides us with a language as well as analytical tools for studying evolutionary processes. A major test is whether Nelson and Winter's evolutionary scheme helps us express such processes as well as reconstruct contributions to the old evolutionary economics.

Nelson and Winter have not been particularly successful in persuading their readers to focus on these kinds of issues, if we may judge from the literature on technical change, theory of the firm, etc. Instead of discussing the Nelson-and-Winter evolutionary research programme as a whole, many readers have chosen to select small parts for citation. In this way Nelson and Winter have been widely cited but their work has only to a small extent been further developed. Nelson and Winter already appear likely to become 'footnote economists', just like Schumpeter. But their work is not very well-suited for this role.

By reflecting upon the character of the task confronted by Nelson and Winter, we may make an *a priori* characterisation of some of the criticism of their results.[61] Non-evolutionary economists may think that evolutionary simulations are unclear ways of reaching results which can be demonstrated with much more clarity and elegance if substantial rationality is not excluded from the analysis. Methodologists may conclude that we have yet another example of the unclear scientific status of evolutionary theory and simulation. Researchers whose individual contributions are included in the synthetic evolutionary process may remark that many of their insights are excluded and that they can easily provide counterexamples against the specific mechanisms which are applied. Representatives of the traditions of the old evolutionary economics may emphasise the primitive character of the model of the evolutionary process which does not include many of their favourite ideas. Statisticians and historians may point out that the quantitative outcome of the simulations relies on arbitrary assumptions and that almost any complex model can be 'calibrated' to fit real data series. Policy-makers may express their disappointment at the lack of clear-cut policy implications.

We could extend this distressing list of types of criticism, but it has little purpose. The real success criterion is whether the creation of a family of synthetic evolutionary-economic processes helps the establishment of a progressive evolutionary-economic research programme (Lakatos, 1970). In other words, we may ask whether Nelson and Winter have helped to create a research programme which increases the understanding of economic evolution and the ability to answer the various criticisms. To me it is clear that in a very general way the answer is 'yes', even if the concrete specifications of Nelson and Winter's models are already beginning to look old-fashioned a decade after the publication of their famous book. The Nelson-and-Winter models have helped others by demonstrating that it is not impossible to treat evolutionary-economic processes in a systematic way. However, it is also clear today that they have not themselves provided a specification of an evolutionary-economic research programme. For example, they have not helped to develop an evolutionary approach to economic transactions (see ch. 4). Furthermore, they have not answered the sceptical comment that their results are simply the outcome of the stochastic character of their models (which is so central to the description of the structure of the search spaces, etc.). Therefore, it is difficult to say whether this element is dispensable or at the very core of evolutionary thinking.

The present emphasis on the basic issues of Nelson and Winter's scheme should not be seen as an attempt to belittle the many examples of applications of simulation exercises that they have provided. It would be nice to have room for discussing the applicability of the Nelson-and-Winter family of models in analyses of economic selection in the manner of Alchian (Nelson and Winter, 1982, ch. 6); factor substitution (ch. 7); economic growth (ch. 9); backwardness and catching-up (ch. 10); general industrial dynamics (chs. 12-14); the role of finance in technical change (ch. 12 and the work of Schuette, 1980); creation of, entry to and exit from an industry (Winter, 1984/91); product change and industrial structure (Gerybadze, 1982). These examples give plenty of material for studying the ways in which the Nelson-and-Winter scheme helps but also sharply focuses (not to say: distorts) the study of evolutionary processes. These examples also allow us to consider how we may adjust their basic scheme and toolbox in order to make them better match the diverse purposes to which they may be applied.

Let us finish this chapter with such an example: market structure or industrial concentration in terms of market shares of the different firms of an industry. Here we see an important difference between evolutionary and structuralist approaches. The structuralists tend to treat the relative market shares as structures, i.e., relatively stable phenomena which help to explain more flexible phenomena (e.g., the conduct of firms and the resulting technological and economic performance). The evolutionists often treat the market 'structures' as relatively flexible patterns to be explained by other factors. Thus, the two approaches often miss each other's points, since what is considered to be the *explanandum* by the one party is the *explanans* of the other party, and *vice versa*.

In this context Nelson and Winter are clearly evolutionists. But what about Schumpeter? Actually, the old Schumpeter appears in his Model II to join a simple structuralist paradigm by developing an argument about the role of 'big business' (Schumpeter, 1942/87, ch. 8). This argument has later been presented as the 'Schumpeter hypothesis' of the superior innovative abilities of large corporations. However, in its crude form this hypothesis should be ascribed to Galbraith who emphasises the importance of firm size *per se* (Galbraith, 1952/72, ch. 7): 'Because development is costly, it follows that it can be carried on only by a firm that has the resources which are associated with considerable size.'[62] But for an evolutionary perspective, the argument seems to be wrongly put: the market structure is just as much a result as a cause of technological development. This point was developed by Nelson and Winter (1982, chs. 13-14) in a series of simulation experiments, but here we will concentrate on the structure of the argument which may help us to rethink the 'Schumpeter hypothesis'. The following short discussion will refer to figure 4.3.

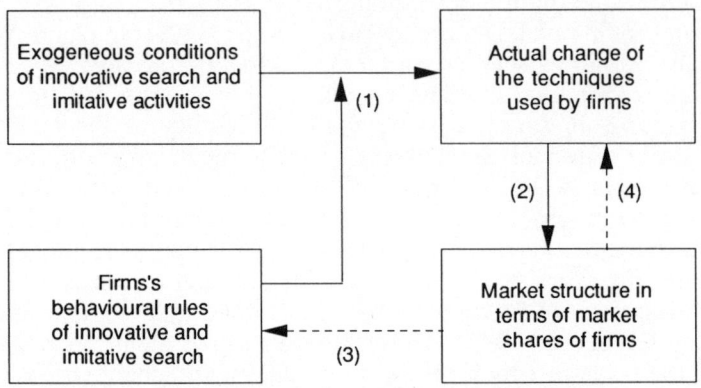

Figure 4.3. Simplified relations between technical change and market structure in Nelson and Winter's and related models.

In the Nelson-and-Winter models discussed above, the chain of causation runs from the exogenous conditions of search as well as the relevant behavioural rules of firms to the technical change of the industry under consideration (arrow 1 of figure 4.3). Given the productivity of firms (and their capital stock), we immediately have the market shares of the individual firms (arrow 2). Here much of Nelson and Winter's work stops. It is, however, possible to create feedback from market structure via rule change (arrow 3) to productivity performance. One story may be that firms with very large market shares have a sharp decrease in motivation for expanding their capacity. For this reason, they may reduce their search costs. The opposite story is that firms with very large market shares want to expand capacity and search costs to create a barrier to entry for outside firms.

In this setting we see that Galbraith was only emphasising the feedback loop (arrow 3). His argument may be rephrased in terms of a minimum level of innovative search costs (economies of scale and scope in R&D). Schumpeter's argument was much broader and includes the elements covered by Nelson and Winter. However, his phrase about 'the perennial gale of creative destruction' (Schumpeter, 1942/87, 87) emphasises that the firms' change of techniques should not always be seen as a gradual process. In this perspective the pace of technical change is increased by competition in a more dramatic way than has been discussed up to now. First, we note that Schumpeter thinks in terms of two modes of technical change and two corresponding types of behavioural rules. In the 'entrepreneurial mode', the search work is performed before entry to the industry (see sections 4.4.2-4.4.3). In the 'routinised-search mode', large corporations succeed in incorporating search into their normal business activities. In this perspective, the growth of firms reaches a threshold which allows a basic change in behavioural rules (arrow 3). Second, we may consider the immediate effect of the large firms (with high productivity) upon the rest of the firms. The latter are simply squeezed out of the industry, and it is not only their capital but also their techniques which are scrapped. Therefore, we see a direct link from market structure to technical change (arrow 4).

This short discussion of the relationship between technical change and market structure has helped to emphasise that we should normally use Nelson and Winter's scheme rather than a concrete example of their models. The same need is demonstrated if we consider Nelson and Winter's treatment of the nature of the search processes underlying the change of rules. With respect to the individual firm it is a question of the amount of resources applied for innovative search for new rules and imitative search with the purpose of copying the rules of more competitive firms. But the incentives to perform search and the probability of being successful depend on the structure, or the 'topography', of the search space. The latter term is used 'to suggest the role of the cognitive conditions under which the search for new methods takes place. The topography of innovation determines what possibilities can be seen from what vantage points, how hard it is to get from one spot in the space of possibilities to another, and so forth.' (Nelson and Winter, 1982, 229)

The explicit consideration of the topography of the search space is an important way of approaching national systems of innovation (Lundvall, 1992). But first we have to connect the search space to economic and political facts. In other words, we want to consider the importance of the nation in terms of the degree of deformation or discontinuity of the search space at the national border. To do so is not easy in the context of Nelson and Winter's concrete modelling exercises since they only deal with search for production techniques in a search space which has little economic structure. This makes it difficult to introduce the idea of national systems of innovation and it even represents a radical delimitation of the concept of 'locality' in the argument of Nelson and Winter, which is especially clear if the search and selection spaces are taken together. To use a metaphor: there are no 'Galapagos Islands' in their topography and thus no chance of developing and testing new variants under specialised conditions before they are tested in the more hostile environment at the mainland or the world market.

The most obvious extension of Nelson and Winter's treatment of the search space is provided by their treatment of imitative search. Even in this case all firms are equally near or far from each other. However, we may distinguish between the probability of imitating firms within the same nation and foreign firms (omitting multinational entreprises). In this way we come to appreciate the coevolutionary path of the firms of a given nationally located industry. If one firm is successful in innovation or in imitating foreign firms, this has a spill-over effect on the other nationally located firms because of the topography of the search space. This idea is relatively easy to implement in terms of evolutionary simulation models.

More complex problems are raised when we turn to the economic and political aspects of the space of innovative search. However, we have a possibility to do so if we combine Nelson and Winter with the idea of interactive innovation (Andersen and Lundvall, 1988; Lundvall, 1992). In relation to the model outline in section 4.4.4, we assume that the capital-producing B-firms have the possibility of producing a large but also largely unknown set of capital goods. The innovative search space is to a large extent determined by the activities of the capital-using A-firms. The behaviour of the A-firms is influenced by the fact that their capital stock puts limits to their change of production routines (and increase of productivity). If the A-firms find new routines, they may not be able to implement them immediately. However, the supply of new types of physical capital from the B-firms allows using these routines. Thus, the search for better routines within the A-firms may lead to the creation of a list of routines which are not feasible in the present period. The access to such lists will radically ease the search conditions of the B-firms but at the same time an equal access of all B-firms to the lists will delimit their relevance for creating competitive advantages. However, there is not equal access to the lists, especially because they do not exist in an easily accessible form; they are difficult to establish and involve a lot of tacit knowledge. But if there is a well-established linkage between two firms belonging to the two industries, there is a large probability that the B-firm will find out about non-feasible routines of industry A. If the B-firm is also active in search of new product-variants, there is some probability that it will discover a machine-type which makes the A-routine feasible. When presented to them, the new machine will be acknowledged by A-firms as having superior characteristics and worth a higher price than ordinary machines.

However, why should A-firms allow the search in their factories for non-feasible routines? If the question should be answered in isolation and if the machine became immediately available to all A-firms, there would be no reason for helping B-firms. But B-firms may be asked whether a new routine can be performed in connection with existing machines and they may be involved in modifying existing machines. Furthermore, a new B-product is normally produced on a small scale in the beginning and the 'linked' A-firm(s) may benefit from early access to the new machines. Such interrelations may cross national boundaries but there are grounds for believing that the probability that they are established as well as the probability that they are successful will depend on national proximity. From a modelling viewpoint these issues are worrying since they greatly complicate the tasks (not only of introducing different vintages of capital but also of creating limited access to the very last machine-type). But in principle they belong to the Nelson and Winter framework.

The two-industry framework is just one of many extensions of Nelson and Winter's conception of innovative and imitative search which allow the discussion of national systems of innovation. The importance of the national setting is considered in terms of the ease of the flow of information or the increased probability of successful products and processes. Once a national economy has embarked upon a trajectory of specialisation strong forces will be reinforcing it. Especially, the search space of the firms located inside this economy has been changed in a way which makes them relatively successful in search related to this specialisation pattern. We may think in terms of specialisation structures of two nations, the 'domestic' economy D and the 'foreign' economy F. The evolutionary argument starts in period t from the behavioural rules and the search conditions which determine the background of the competitiveness of firms and industries. Via the actual technological development of the industries (e.g., A and B), a specialisation pattern emerges which can be measured in terms of concentration or export specialisation. However, in period $t + 1$ the specialisation structure and the related technological conditions will function as the starting point for considerations about change in the behavioural rules. Furthermore, the actual technological development in some nationally located firms and industries in one period will influence the conditions of search for other firms in the next period.

To conclude, we should note that it is through modifications of the Nelson-and-Winter framework like the ones just discussed that we are placed in a position which allows an appreciation of it. However, there is an obvious possibility that Nelson and Winter become established as 'footnote economists' who are frequently cited but not considered as related to a viable research programme. The present introduction to parts of their work is intended to suggest some ways of moving their framework from the footnotes to the main text. In particular, the programming appendix suggests explicitly and implicitly a number of tasks for future research as well as a powerful way of communicating about evolutionary mechanisms and processes.

Notes

1. In this chapter 'Nelson and Winter' is used to denote works on evolutionary modelling with joint authorship and a couple of follow-up works with individual authorship which clearly relate to their book (1982). The models presented here are called 'Nelson-and-Winter models'. There should be no difficulty in distinguishing these common works from the individual works of the two authors.
2. The notion of 'routines' suggests a certain explanation of evolvable knowledge (see Nelson and Winter, chs. 4-5). In the present account I shall apply the more general notion of 'rules' which also opens the way for other types of explanation.
3. To avoid confusion about the concept of complexity it is important to distinguish between three different types of complexity: (1) data complexity (or entropic complexity) of the type discussed above, (2) problem complexity (or logical depth) which deals with the number of computational steps needed by an algorithm to solve a problem (dealt with in the present section), and (3) system complexity (or organisational complexity) which has also been mentioned above. As an example of the difficulties which emerge from a too little developed concept of complexity and an imprecise relation of it to evolutionary processes we may take the ambitious and inspiring but also confusing book by Nørretranders (1991) on the limits of an intellectualistic approach to the understanding of human consciousness and capability.
4. The algorithmic approach emphasises the processes within the economic system rather than the results of the economic process. To select between the (infinitely) many different algorithms which can generate a certain set of observed results, we have to define a criterion. Here evolutionary economics must make a difficult combination of realism and simplicity.
5. Or: Good Old Fashioned Artificial Intelligence as opposed to commercial expert systems, etc., see Haugeland (1985/89, 112-117, 176-185).
6. My emphasis on Simon is not meant to suggest that he developed all the ideas. In this respect, it would have been more fair to write Newell and Simon (see Simon, 1991a, ch. 13).
7. The ideas of substantive and procedural rationality were present in Simon's thinking long before they got their names (in Simon, 1976/82; 1978/82).
8. This last item is implied by the former items and is not mentioned in Simon's list. However, this redundant item achieves its meaning when the list is negated.
9. To use the notion of Hayek.
10. Which is, unfortunately, spread over several of his works, see Simon (1981; 1982). Simon's argument for rule-based satisficing behaviour as a means of complexity reduction is summed up by Heiner (1988). In this formulation rule-based behaviour will prevail if there exists a large gap between decision-makers' competence in using information and the difficulty of their decision problems. The more complex the situation, the more rational it is to follow rules. Further arguments for rule governed behaviour have been developed by game theory (Kreps, 1990a, 65 ff.). For example, we may note the argument that if a firm is (for a period) able to make an irrevocable commitment to a given decision rule, then this rule will influence the outcome of the game. Such commitments from the side of a monopolist may deter potential competitors from entry into the industry. On the other hand, we see that Simon's general case for bounded rationality and retrospection appears to be of much interest to modern game theorists (see Kreps, 1990a, ch. 6).
11. The interconnections between Simon's manifold studies in management science, economics, psychology, epistemology and computer science are described in Simon (1991a).
12. From another perspective a somewhat similar list is suggested by Flores and Winograd (1986).
13. The attempts by Simon to move from Artificial Intelligence and cognitive science (see Newell and Simon, 1976) back to economics (see Simon, 1982, part 8; 1986; 1991a) are mainly characterised by his promotion of earlier ideas.

14 For the difference between phylogenetic evolution and ontogenetic evolution (which is now called development by biologists), see Gould (1977) and Hodgson (1993, ch. 3).

15 This is, of course, a stylised version of the background of the family of models created by Nelson and Winter. They themselves give generous acknowledgements to these and other 'allies and antecedents of evolutionary theory' (Nelson and Winter, 1982, 33-45). In their initial account (ch. 2) they have forgotten American Institutionalism which is, however, mentioned in passing on p. 404.

16 The readiness of Simon's framework for an evolutionary interpretation is demonstrated by the fact that he had no difficulty in his later integration of Nelson and Winter's ideas into his own framework, see Simon (1981, 52-57; 1983, ch. 2).

17 The different chapters of Nelson and Winter's book (1982) include the following models (named after their respective chapters): Model 6: 'A Particular Model of Economic Selection' (pp. 144-154); Model 7: 'A Markov Model of Factor Substitution' (pp. 175-184); Model 9: 'An Evolutionary Model of Economic Growth' (pp. 209-214; see Nelson et al., 1976); Model 10a: 'Development and Backwardness in a Two-Technology Evolutionary Model' (pp. 235-240); Model 10b: 'Many Techniques and Many Variable Inputs' (pp. 240-245); Model 12: 'Dynamic Competition and Technical Progress' (pp. 281-287, 302 f.; see also Nelson and Winter, 1978, 544-547). Other researchers, like Silverberg et al. (1988), have contributed with related evolutionary models which, however, cannot be dealt with in the present context.

18 More specifically, we are dealing with a generic model with two major variants. One variant is Model 9 (see the previous footnote) which is probably the most well-known of the Nelson-and-Winter models, since it was used to demonstrate that evolutionary models can generate aggregate growth and productivity data (for the period 1909-1949) just as well as neoclassical growth theory but with more 'realistic' microfoundations. This variant deals with a set of firms producing a homogeneous output with techniques which vary with respect to capital and labour productivity and which firms search for when they have unsatisfactory results. The other variant is Model 12 which is designed to study 'Schumpeterian competition' in an industry with a homogeneous product but with techniques which vary with respect to capital productivity. In the subvariant Model 12a (Nelson and Winter, 1982, chs. 12-14) the search is independent of the profitability of the firms. In subvariant Model 12b (see Winter, 1984/91, 284 f.) we find adaptive change of search policies according to a satisficing mechanism. In subvariant Model 12c entry to the industry and industry birth is introduced (Winter, 1984/91, 285-288). The presentation will concentrate on Model 12a which is the most well-described of the Nelson-and-Winter models. Model 12b and 12c are covered in section 4.4.

19 The computational flow chart of figure 4.1 does not describe the constants of the model, including the exogenous cost and market conditions, firms' capacity utilisation rules, behavioural rules of banks, the disembodied character of technical change, the structure of the search space in terms of search costs and productivity of (probabilistic) search results. These elements are included in the formal version of the model in section 4.3.

20 The only attempt that I know of to describe Nelson-and-Winter Model 12 by a flow chart of the related computer program is found in a published dissertation by Gerybadze who also uses this technique to express his own variations of the Nelson-and-Winter model, see Gerybadze (1982, 129, 228 f., 301 f., 388 f.). Figure 4.1 is intended to reflect the computational structure of a Nelson-and-Winter model more precisely than Gerybadze, 1982, 129. Some aspects of Nelson-and-Winter Model 9 are depicted by a flow chart in Coombs et al. (1987, 117).

21 There are, of course, other possibilities. In other examples of this generic Nelson-and-Winter model (see section 4.3) satisficing behaviour is applied and labour productivity as well as capital productivity is taken into account, see Nelson and Winter (1982, 209-214) and Nelson et al. (1976).

140 *Evolutionary Economics*

22 The word is used by Nelson and Winter (1982, 285). It is perhaps not the best word, since 'draw' often means a 'tie' where no one wins.
23 Nelson and Winter have made the financial constraint dependent upon current profitability and have neglected balance sheet magnitudes in order to avoid a large set of interrelated complexities which would hide the main ideas of their simple models, see Winter (1984/91, 283).
24 The FORTRAN programs of Nelson and Winter originate in the period before these programming paradigms were applied (see the program listings in Schuette, 1980). This may to some extent explain the lock in of the trajectory of Nelson-and-Winter models into a relatively limited set of assumptions, functional specifications and parameter values. It also means that the transformation of their formulations and results is not at all easy. In the appendix to this book, I have abstained from a direct application of a fully modular approach with abstract data types. Most of the appendix is only following the principles of structured programming while a fully modular approach (combined with the possibilities of object-oriented programming) is only sketched very briefly.
25 But other frames of reference—such as evolutionary-biological modelling of replicator dynamics—are, of course, also available.
26 See Nelson and Winter (1978, 155-157; 1982, 284-286, 302-303) and Winter (1984/91, 281 f.). My restatement uses in part a modified version of the imperative computer programming languages, see the appendix to this book.
27 Nelson and Winter's lack of interest in a systematic exposition of the details of their simulation models is most obvious with respect to Model 9 (see Nelson and Winter, 1974; Nelson et al., 1976; Nelson and Winter, 1982, ch. 9). With respect to Model 12, a marked improvement of the exposition is given (see Nelson and Winter, 1978, 544-546; Nelson and Winter, 1982, 284-286, 302-303). However, much is still lacking in order to teach the readers how to use the Nelson-and-Winter tool-kit. Some of it is provided by Winter (1984/91).
28 Like $\theta^n(\ )$ and $\psi(\ )$ of statements (4.10)–(4.11).
29 If we had concentrated on Nelson-and-Winter Model 9 rather than Model 12a, it would have been necessary to compare it with neoclassical and post-Keynesian growth theory. Since the Nelson-and-Winter models started by formulating an alternative to growth theory (Nelson and Winter, 1974), it is not without interest to study how their initial reformulation of the way capital productivity is changed in the neoclassical growth models has influenced the further development of the class of Nelson-and-Winter models. However, we have to exclude this issue from the present exposition.
30 Observe that c is the same for all firms and all periods.
31 The 'draws' are interpreted as the outcome of a discrete binomial approximation to a continuous time Poisson process. The smallness of the probability of having a 'draw' secures a good approximation.
32 The outcome of the innovative 'draw' is determined by a distribution of technological opportunities which depend on time and on the current productivity of the firm (see Nelson and Winter, 1982, 302). $\log(A_{it}^n)$ has a normal distribution with a time-dependent mean $A^{init} + \varphi t$ and a standard deviation σ^n. For further specification see appendix (A.2.6).
33 This subject raises several difficulties which are treated in the procedures of the appendix (A.2.7).
34 $G(\ )$ is specified in a way which secures that in case $\pi_{it} \leq 0$ there is no external financing. Otherwise external finance is given in proportion, b, to the economic profit per unit of capital. Depreciation is somewhat surprisingly introduced into the calculations of the financial constraint by Nelson and Winter (1982, 303). For further specification see appendix (A.2.7).

35 Some errors in the components of this equation in the cloth version of Nelson and Winter (1982, 285 f.) have been corrected in its paperback version without any other indication of a difference between the two versions.
36 For a description of the area of Artificial Intelligence, see e.g. Rich, 1983. The most notorious critique of Artificial Intelligence is Dreyfus, 1972/79. The beginnings of a critique of Artificial Intelligence from an evolutionary perspective (in the manner of Manturana) can be glimpsed in Flores and Winograd, 1986.
37 The importance of such developments is described by one of the founding fathers of Artificial Intelligence in Simon, 1991a.
38 However, we may also expect something like 'overselling' of premature results which has been a constant factor in the history of Artificial Intelligence and not least in Simon's promotion of it, cf. Dreyfus, 1972/79, 81-85, 93-96.
39 The search space is randomly generated and not the same as in section A.2.1.
40 Only cases building directly on the basic Nelson-and-Winter model are taken into account. This means, e.g., that evolutionary models with a vintage structure of capital, like Iwai (1984/93; 1984) and Silverberg et al. (1988) are not covered.
41 This and the next section (4.4.2) present extensions of the basic Nelson-and-Winter model build on Winter, 1984/91, which is reproduced in Winter, 1986, where we also find a discussion with contributions by Eliasson, Ungern-Sternberg and Winter, see Day and Eliasson (1986, 233-245).
42 We should notice an important difference between Winter's performance indicator and the aspiration level in Simon's model of satisficing behaviour. The latter is fixed while the former adapts itself to past experiences with respect to the profit rate. With respect to Winter's formulations, one may ask how the firm can get immediate access to this piece of statistical information as well as the average search efforts ($\bar{r}_t^n$ and $\bar{r}_t^m$). Some lags and errors are likely to occur. However, instead of complicating the analysis, we shall here accept Winter's version of the story.
43 With respect to the sequence of computations, the following assignment statements fit well after the foregoing (4.1)–(4.22).
44 With standard deviations σ^n and σ^m.
45 See Winter (1984/91, 285). Here we also find several other features to make the satisficing story more realistic.
46 The idea is developed by, e.g., Freeman et al. (1982, 35-43).
47 Because of the model structure, the consequence of this is that the firm is liquidated at the end of period t and that it will never emerge again. We assume that the physical capital of the firm, K_{it}, is scrapped rather than reused by other firms. For simplicity we allow the firm's name (its identification number) to exist forever. Because of this the meaning of n changes to the number of firms which have ever existed, n_t^{id}, and zeros will exist in the state vectors K_t, A_t, r_t^n, r_t^m and X_t.
48 There is no simple way to describe this search by industry-external actors in terms of satisficing behaviour (as in (4.24)–(4.25)).
49 Statement (4.26) as well as the following statements ((4.27)–(4.32)) may be placed in the computational sequence after (4.1)–(4.25). We are clearly approaching a point where conventional programming concepts like arrays are not really adequate. Instead we may turn to object oriented programming which combines local data structures and local procedures to 'objects'. A firm may be such an object. The actor creating a firm is quite another type of object who has the power of allocating the real-world analogues to variables implemented in a computer memory (see appendix, section A.3).
50 The formulation of this series of statements is made as similar to (4.29) as possible. It is easy to introduce more differences between the innovative and imitative entry to industry.
51 With a standard deviation, σ^e.

52 However, the initialisation procedure of (4.29) must be modified since it builds in industry averages.
53 The discussion here is based on the model rather than on simulations. In Winter (1984/91, 291-298) we have a discussion which is solely based on a comparison of industry creation by an entrepreneurial regime (Schumpeter Mark I) and a routinised regime (Schumpeter Mark II). Because of this comparison, Winter has little occasion to reflect on the general features of the evolutionary process.
54 A discussion of the long-term trend in the size structure of the industry is omitted since it depends on specific assumptions about the numerical values of the model.
55 In this context the effect of a government quality control (Rule IV) may be described as a modified breakdown vector. There is no reason to believe that the entries in this vector are changed to zero. To see this it is useful to consider different types of breakdowns with individual and independent probabilities. Government intervention only concerns a few of the breakdown types whose probabilities are at best reduced nearly to zero. However, most of the (less drastic) breakdown types have unchanged probabilities. The result is probably that there are still differences in the quality of the brands but that the ranking of the brands has been changed.
56 Later we will see that the search behaviour can be reintroduced in industry A.
57 In practice this assumption does not hold because of, e.g. (1) lack of general information about new capital goods, (2) lack of production-specific knowledge which can only be obtained by a prolonged period of learning by using and learning by interacting with suppliers, (3) the incremental innovations which take place within each type of capital good. In principle, such factors may be taken into account but only at the cost of a radical increase in the complexity of the model. As mentioned earlier, the simplicity of the models is a means of avoiding many of the well-known pitfalls which are connected with computer simulation by means of huge models.
58 See (4.11).
59 See (4.14).
60 We do not need an explicit identification of A-firms in the present shortened exposition.
61 Practically all the types of criticism (as well as several others) are found in the review by Mirowski (1983) of Nelson and Winter's book.
62 In this version the hypothesis has created a huge and confusing literature reviewed by Kamien and Schwarz (1982) as well as by Dasgupta and Stiglitz (1980).

5. The evolution of strategies of transaction: an algorithmic story

5.1. Rule change in iterated computations

5.1.1. Rule-based systems and genetic algorithms

To cope with a large number of different agents, it is necessary to abstract from the processes of individual decision-making or at least to consider them in a highly simplified form. This necessity to simplify was confronted by Alchian (see section 1.1.3) in his example of a population of travellers setting out from Chicago: thousands of travellers without any foresight explore an environment where the selection criterion is the availability of gasoline stations. This example is easily translatable into the terms of heuristic problem solving (section 4.1.2), but it is not an individual decision-maker who performs the problem solving. The problem of finding a feasible route is solved by the whole population of travellers, and this happens in an extremely wasteful way: most of the travellers have to walk back to Chicago but some solve the problem of getting away from the city. The underlying heuristic of trial-and-error is clear enough: try everything; some will succeed.

The possibility of applying a population-oriented approach to heuristic problem solving was observed by the pioneers of Artificial Intelligence, and this may explain why Simon appears to have been able to embrace Darwinian evolution (Simon, 1981, 200-209) as well as the new evolutionary economics of Alchian, Nelson and Winter, and others (Simon, 1981, 52-57) in his own terms on heuristic problem solving:

> The simplest scheme of evolution is one that depends on two processes; a generator and a test. The task of the generator is to produce variety, new forms that have not existed previously, whereas the task of the test is to cull out the newly generated forms so that only those that are well fitted to the environment will survive. In modern biological Darwinism genetic mutation is the generator, natural selection the test. ... [However,] Winter and Nelson observe that in economic evolution, in contrast to biological evolution, successful algorithms may be borrowed by one form from another. Thus the hypothesized system is Lamarckian ... (Simon, 1981, 52, 57)

The application of the analogy between trial-and-error search and a Darwinian-type evolutionary process should not hide the fact that Simon has shown very little interest in evolutionary analysis. There appears to be a very large step from his version of Artificial Intelligence to new evolutionary economics. One problem is his sole focus on the individual decision-maker placed in a complex environment rather than a heterogeneous population of interacting decision-makers.[1] Other Artificial Intelligence researchers have shown more interest in changing their perspectives and the issue of evolutionary problem-solving is much discussed at present. However, the issue is still marked by some confusion about how to characterise the outcome of the evolutionary heuristics of problem-solving: is the result an optimising one as Friedman (1953) suggested in his reinterpretation of Alchian, or is it really wrong to characterise the outcome of a process of selection as optimisation, as argued by Winter (1964; 1971/93; 1975).

The results of the theory of complex problems give a foundation for advancing this discussion in a much more precise way than hitherto. Especially, the difference between local and global maxima has become much clearer than previously. Darwinian evolution of species proceeds gradualistically in a landscape mapping the relationship between hereditary characteristics and reproductive success. A species cannot descend from one hilltop of fitness in order to reach an even higher hilltop; but real evolution takes place in a multi-dimensional landscape and here are many more evolutionary possibilities in the form of saddle-points, contour lines, etc. (Küppers, 1986/90, 101 ff.) However, there are still many limitations to the evolutionary process which are only partially overcome by allowing some degree of strategic behaviour into the system. As long as

... we are considering a system that lives in an environment having a multitude of local maxima, we cannot understand the system or predict its behavior unless we know something of the method and history of its evolution. Nor can we judge whether there is any reasonable sense in which such a system can be regarded as 'fittest'. (Simon, 1981, 55)

This conclusion has also been drawn by researchers dealing with, e.g., game theory (Kreps, 1990a, ch. 6) and Artificial Adaptive Agents (Holland and Miller, 1991). The conclusion sets the scene for studies which try to mimic the real mechanisms and the realistic behavioural parameters (Arthur, 1991; 1993) of an evolutionary process. At present the problem is not a lack of supply of suitable algorithms and model schemes for solving this task as it was when Schumpeter had his tool-problem (section 1.1.2). On the contrary, we have to avoid an unsystematic application of a multitude of approaches, or of individual tools within a specific approach. For example, the Artificial Adaptive Agents (AAA) used to model evolutionary processes allow a wide range of computer-based adaptive algorithms. 'Usually, there is only one way to be fully rational, but there are many ways to be less rational. It is important in building a theory based on AAA to construct agents that exhibit robust behavior across algorithmic choices.' (Holland and Miller, 1991, 367)

When trying to create a population of heterogeneous agents which changes in a way which is both adaptive and creative, we have great problems in using traditional programming approaches for modelling the individual decision-maker or firm. We might, of course, develop the work of Cyert and March (1963) and the many others who have dealt with computer simulations of firms in the last thirty years. But even with vastly more sophisticated computers we rapidly reach a level of complexity where confusion sets in. Therefore, adequate conceptual and programming tools are as important as they ever were. A simple but powerful tool may be used to discuss the possibilities of introducing a population perspective into our analyses. This tool is called a rule-based system or production system and it was originally invented by the logician Post who demonstrated that it is one of the models of computation which is equivalent to Turing machines. It was developed into a tool of Artificial Intelligence by Newell[2] and it is now well-known as a component in expert systems. The question is now whether the tool is of any use to the new evolutionary economics.

In informal terms we can say that a formal production system consists of:[3]

1. A set of 'productions' or rules of the form: ⟨condition⟩ → ⟨action⟩.
2. A 'workplace' or database that provides input to the rules and that is modified by the actions of the rules.
3. A control unit which especially resolves the conflicts which arise when more than one rule is applicable at the same time.

The core of a production system is the set of rules which may as a first approximation be interpreted as a set of primitive economic agents who all behave according to the stimulus-response model of behaviouralist psychology: when an agent meets his particular signal, he acts in a pre-programmed way, just as Pavlov's dogs. This is, of course, only one of many interpretations. The underlying procedure is purely formal and has as such no meaning whatsoever. But let us stick to the interpretation in terms of economic agents.

The rule-based approach emphasises the modularity of the production system: each rule or agent can be considered as a module with a well-defined way of functioning in relation to its environment (Rich, 1983, 32; Haugeland, 1985/89, 162). The rule is not invoked explicitly as in procedural programming languages; instead the system is 'data driven' in the sense that any agent who finds his condition fulfilled in the working space will try to act. Therefore, the system has a large degree of implicit parallelism and this has been exploited in modern parallel computer systems. Another characteristic is that it is easy to modify existing agents or to introduce new agents into the set of rules without revising other parts of the system. This modification may even be performed during a computation, and in this way we approach an evolutionary process in which irreversible change occurs.

An important development with respect to rule-based systems was made by Holland's in his book on *Adaptation in Natural and Artificial*

Systems (1975). This book started the development of a systematic approach to adaptation, optimisation and learning, an approach which combines two major tools: classifier systems and genetic algorithms. Let us shortly consider these tools:

A classifier may be interpreted as a behavioural rule of an agent, and thus it consists of a conditional part and an action part. Its special characteristic is that the conditional part is formulated in terms of the set of situations (or 'messages') that satisfy it (see Holland *et al.*, 1986, ch. 4). The background for this approach is that even if the situations in which the agent act may be described as bit-strings of length λ, the agent does not need to take all this information into account. If the agent has been exchanging goods with another agent λ times, and if the quality of the goods received each time is characterised as 'perfect' or 'defective' ('1' or '0'), then the information available for the next move of the agent is a list of length λ, describing the qualities of received goods. For example, 1000 means that defective goods have been received in the last trading round while perfect goods were received in the three previous rounds. The agent will normally only take into account a little of this information. This means that his strategy should include the possibility of a 'don't care', which is symbolised by '#'. If the conditional part of his rule is $\overline{1\#\#\#}$,[4] the rule is applied if the quality of the last shipment of goods was perfect; the quality of all previously received goods is ignored,

In the beginning of each round of the computation, each agent checks whether his condition is fulfilled by the data present in the workspace of the computer memory. If it is not fulfilled, he remains passive in this round of the computation; if the condition is fulfilled, then he has to queue up together with all the other agents who have got their right stimulus. When all agents have made their decision, the control unit checks whether there is a competition between some agents who want to perform the same action. This competition is decided by means of a measure of the relative strength or competitiveness of each agent. The one with the highest score gets the task (or has a higher probability of getting the task). Then the control unit revises the competitiveness indices of all agents in a way which increases the relative strength of the ones who have been allowed to be active. Finally, the agents do their 'productive' work in the workspace, and we are ready for a new round of computation.

A series of computational rounds determines success and failure of the agents (and thus of the behavioural rules) of the classifier system. But this computational process does not investigate the competitiveness of the rules which are not present in the computer memory. The problem is that the situational information is so massive, that only a tiny part of the possible behavioural rules can be tried out. It is to cope with this problem that Holland introduces genetic algorithms for changing the conditional part of the rules. This change is explicitly modelled after the genetic mechanisms of sexual reproduction, like mutation, crossover and inversion of the bit-strings defining the behaviour of the agents. This means firstly, that after each round of computation, agents are allowed to multiply according to their competitiveness indices. In the next round, successful agents are copied, while unsuccessful agents are removed. Secondly, in each round, the 'genetic' composition of agents (the conditional part of their rules) is changed with a small probability. A point mutation of the above-mentioned condition for the behaviour of a trader may, e.g., lead to the condition $\overline{1\#0\#}$. This new condition means that the agent will only react if the quality of the goods delivered in the last round was perfect while the quality of the goods delivered in the third-last round was defective. Thirdly, the evolutionary process is allowed to take place over a large number of computational rounds. During this process, the average performance of the agents tends to increase for a period and then to stabilise. We cannot be sure that a global optimum is ultimately found, but, nevertheless, the system has been shown to be a good tool for coping with complex optimisation problems (like the ones in section 4.1.2).

Through the work on classifier systems and genetic algorithms, we now have a well-developed approach to the subdiscipline of Artificial Intelligence which is called machine learning (Holland *et al.*, 1986; Davis, 1987; Booker *et al.*, 1990).[5] In the present context, the question is whether we also have an important approach to evolutionary-economic modelling. Probably it is still too early to answer this question, especially because the introduction of genetic algorithms and classifier systems to economics is still influenced by their origin in the attempts of Artificial Intelligence to develop heuristics on complex optimisation problems (Holland and Miller, 1991). However, the tools have also in promising ways been applied within core areas of evolutionary-economic modelling.

5.1.2. Arthur's application of classifier systems

We shall not try to start by a full-blown application of Holland's version of the evolution of rule-based system. Instead we shall show how a partial Holland-type model allows us to account for a relatively realistic process of rule change of an individual agent and show how such rule changes can be given a learning-by-doing interpretation. The best example has been given by Arthur (1991, 353) who assumes that 'human rationality is bounded. The question is how to model economic choices made under these limits.' He suggests an experimental approach where artificial economic agents are constructed and parameterised in a way which ideally should allow them to pass the Turing test by being exchangeable with real human agents in teleprinter dialogues (p. 354; 1993, 3). In reality I will suggest that Arthur does not follow Turing's goal; instead he deals with artificial agents in a closed and experimental problem situation. Within these limits, the task is not to make artificial agents as effective as possible in an iterated decision situation but 'to design a learning algorithm or learning automaton that can be tuned to choose actions in this iterated choice situation the way humans would.' (1991, 354) The question is which kind of algorithms are most relevant for this purpose; the answer is that even if he does not believe in the existence of a generally applicable decision algorithm,[6] he finds classifier systems combined with genetic algorithms useful in situations where actions may be improved by using the information feedback from the environment, but where there is also room for exploration of poorly understood actions. However, he only uses a small fraction of this framework, especially classifier systems.

Arthur describes a learning automaton or an artificial agent who among n different possibilities chooses one action in each period. The agent bases his decision on past experiences of the payoff of different actions and after each action he updates his probabilities of taking the different actions on the basis of the newly won experience. In each period the same things happen but the agent's confidence in his decisions increases and as time goes on he becomes more and more unwilling to change his preferred action. In the end he sticks to one specific alternative. The question is now: is it the optimal one? This depends on the parameter of confidence-building.[7] If confidence-building is too fast, then there is a danger that the agent's behaviour will lock-in to a possibility which is non-optimal; a relatively good solution that is found in an early period will be preferred to an uncertain exploration of the space of possibilities. If confidence-building is relatively slow, then the behaviour will end up close to the global optimum.

In relation to Arthur's model, we may explore more active forms of search for a high-quality brand than covered by the simple Markov model with brand loyalty except after an experienced breakdown (section 4.4.4). Actually, the model of an artificial adaptive buyer presupposes that each buyer has a record of his perception of the 'strength' of each brand and that he experiments with different brands for a shorter or longer period. It supposes that there exist important qualities which can to some extent be considered as a characteristic of each brand. It is, however, a stochastic quality which can only be learned imperfectly through 'probability learning'.

In such a situation, a single trial of a brand may give a wrong impression because of the dispersion around the mean value. For such a situation the behavioural rules of section 4.4.4 are clearly inadequate. The method of market-share based choices is also dubious in the early phases of the evolution of the system. Instead we may explore a heuristic which helps the buyer to find for himself a brand which has, or comes close to, the maximum quality. Such a heuristic includes four rules:

   I. Keep a record of the perceived strengths of the different brands.
   II. Update this record based on experiences with the different brands.

III. Decide brand selection in a probabilistic way where the probability of choosing a particular brand is deducted from the record of strengths.
IV. Determine a confidence mechanism which ensures that experimentation stops when there is reason to believe that a brand has been selected which has, or comes close, to the maximum performance.

A very good example of a heuristic which includes rules I-IV is defined by the following algorithm:[8]

At the beginning of each period, t, do the following:
1. Add the measure of experienced performance of the brand used during period $t-1$ to the already existing strength measure of that brand, $S_{i,t-1}$.
2. Renormalise all strength measures so that their sum becomes equal to a parameter β, i.e. $\sum_{i=1}^{n} S_{it} = \beta$.
3. Define the probability of choosing brand i for period t to be $p_{it} = S_{it}/\beta$.
4. Select a unit of a brand based on these probabilities.

This algorithm ensures both a period of experimentation and an absorbing state where one brand is always selected by each agent. The relationship between parameter β and the strength measure of an individual product evaluation determines how much time is given to learn about the qualities of brands before the final decision is made. The larger the difference is between the value of β and the values of the quality performance of brand units, the less emphasis is put on an individual result and the more time is given before one brand obtains a strength equal to β and thus a probability of being selected equal to 1.[9] Before this absorbing state is reached, there is a period where the buyer's confidence in his favoured brand choice increases and where he becomes more and more unwilling to change his preferred action. The fast confidence-building may lock the agent's behaviour to a possibility which is not the best. The slow confidence-building makes possible a brand choice close to the optimum.

Arthur's studies appear to indicate that artificial agents can be 'calibrated' to behave closely to humans in similar games. The estimated parameter of confidence-building (learning) appears, furthermore, to be insufficient to secure a close fit to the optimal solution.[10] However, the answer will probably differ strongly between different types of goods and different parts of economic life. If there is a characteristic learning time (or confidence-building time) in different parts of economic life, it becomes very important to compare it to the time-span over which there is a relative constancy of the economic decision problems in a more or less turbulent economic environment. (Arthur, 1993, 17)

These and other issues are raised by the study of the artificial economic agents. But in the present context we are more interested in the question whether a set of Arthur's agents can demonstrate the functioning of a full evolutionary process. The answer is clearly no as far as the model presented is concerned. The reason is not that it has a Lamarckian rather than a Darwinian flavour. The reason is simply that Arthur's artificial agents are still exploring a given universe by means of given rules. Such a process will always end up with all the agents bound to a particular (optimal or sub-optimal) behaviour. In other words, the possibility of a persistent creation of new behavioural variety to deal with the ever-changing economic environment (which to a large extent consists of other agents' behaviour) has not yet been dealt with.

However, Arthur's leaning to the work of Holland serves to suggest ways of ensuring new variety of rules as well as for creating a complex environment which may open up new niches and allow the gradual emergence of systemic complexity (Holland and Miller, 1991). Even in this new context, Arthur's question about characteristic learning times is, however, quite central. Any evolutionary process will come to an end if one of its mechanisms becomes too dominant: if confidence-creation or norm-guided behaviour dominates, a sub-optimal stasis may develop; if the selection of the fittest behaviour is very strong, a stasis will emerge which is more optimal with respect to a static environment; if variety-creation comes to dominate, the outcome may, however, be chaotic.[11]

5.1.3. From production routines to rules of interaction

Given the emphasis in the present book on Nelson-and-Winter models, it would have been very helpful if it had been possible in a simple and faithful way to transform such models into the language of classifier systems and genetic algorithms. This is, unfortunately, not possible. Instead the primary area of application of these tools are presently the evolution of strategies within iterated games.[12] A pioneer in this work is the political scientist Axelrod who by his emphasis on conflict and strategy comes closer to much economic thinking than Nelson and Winter's models, which take their starting point in organisation and technical change. This difference is based on different answers to the question 'What evolves?' Nelson and Winter's answer is apparently 'organisational routines in general', but a closer look reveals that only a certain kind of routine is taken into account.[13] The evolving routines are technical, and the search is only related to economic variables on the cost side. Axelrod's answer is that what evolves are strategies of interaction. This means that his models introduce information about the interaction between actors during a series of periods. By looking backwards at the results of their previous interactions, the actors may try to guess which strategy will be successful in the future. Successful guesses are more likely to increase in frequency, but they, at the same time, change the environment for other strategiesAs a result, the frequencies of strategies evolve in the long-term perspective.

The conceptions of the Nelson-and-Winter and Axelrod types of evolutionary model are so different that the possibility of a synthesis between them has not yet been discussed. To simplify, we may think of Nelson-and-Winter models as focusing on organisations and their production techniques while Axelrod-type models may be developed in a way (see below) which focuses on economic interactions, not least transactions. This is one dimension in the two-by-two scheme in table 5.1. The other dimension concerns the type of rationality implied by the models: substantial (unbounded) and bounded rationality. While mainstream microeconomics (even of the game-theoretical kind) still sticks to unbounded rationality with very strong assumptions about the kinds of information and computational capabilities available for the economic agents, all types of evolutionary modelling consider rationality to be bounded in Simon's sense.

Table 5.1. Four types of paradigms in the analysis of economic change.[14]

Rationality viewed as:	Focal concern:	
	Organisation/production	Interaction/exchange
Unbounded	(1)	(2)
Bounded	(3)	(4)

Like most simple taxonomies, the present one suggests an oversimplification. However, it serves to expose the general similarities and differences between Nelson and Winter's and Axelrod's types of model. Nelson and Winter agree with Simon (1991b) that the understanding of organisations rather than of markets is the starting point and major aim of the analysis. On the other hand we have Axelrod whose main contribution explores the best known of all the games used in game theory: the Prisoner's Dilemma. Here the characteristics of organisations/players are only considered to the extent that they are of interest to the problem of interdependent decision-making. Axelrod has dedicated several studies (see Axelrod, 1984; Axelrod and Dion, 1988) to the iterated version of the Prisoner's Dilemma game.

If we reinterpret and develop these studies, we have an example of an evolutionary process which includes notions like uncertainty, procedural behaviour and its payoffs, competitiveness, imitation and innovation, observed and theoretical fitness, coevolution, stasis, mass extinctions, etc. These notions are clearly developed from the viewpoint of transaction. The case is relatively complex (and some of its details may be ignored during a first reading) but in return the case gives an introduction to central aspects of the new evolutionary economics which are not covered by the Nelson-and-Winter models.

The present emphasis on an algorithmic approach to Axelrod-type models determines the kind of details that will be treated. We shall concentrate on the development and implementation of a 'story' around the iterated Prisoner's Dilemma rather on reviewing and commenting upon the huge literature on the subject (not least related to the concept of evolutionary stable strategies; see below). This means that many interesting developments cannot be included in the exposition, for instance, the related work on the role of institutions in the revival of trade (Milgrom et al., 1991). In this work it is demonstrated how a large number of iterated Prisoner's Dilemma games can be transformed into simple coordination games. In relation to my development of the story of the so-called Trader's Dilemma, it should, however, be remarked that Milgrom et al. relate their theoretical results to the emergence of private judges and rules for participating in the Champaigne Fairs of the early Middle Ages.

5.2. From Axelrod to the Trader's Dilemma

5.2.1. Axelrod's evolution of cooperation

Axelrod's starting point is simple: suppose the situation depicted by the Prisoner's Dilemma is not a once-and-for-all confrontation of the two[15] actors involved. Instead the playing of the game is repeated or iterated a number of times with the same two players involved. Can this change in the rules of the game help us to understand how defective behaviour is transformed into cooperative behaviour? Axelrod's answer is 'yes' and his method is computer simulation of a population of players repeatedly engaged in Prisoner's Dilemma interactions. In this context, he tries to demonstrate that even a non-cooperative starting point may allow for the evolution of cooperative behaviour.

The game he discusses in his book on *The Evolution of Cooperation* (Axelrod, 1984) as well as in several papers (summed up in Axelrod and Dion, 1988) is very different from the single playing of the Prisoner's Dilemma. The latter may be interpreted in terms of a situation created by the criminal authorities to lure Bonnie and Clyde to betray each other. It is a situation where rational decision-making calls for defective behaviour, even if cooperative behaviour would make both players (prisoners) better off. According to Axelrod, the situation depicted by the Prisoner's Dilemma will better reflect social problems if it is thought of as an iterated game. Two duopolists may be involved in repeated price wars if they do not find a tacit form of cooperation. Two merchants may have to make repeated barter trade with each other, and thus they have to decide repeatedly whether to cheat or not. Two nation states may have repeated trade wars.

This kind of situation has been much studied by analytical game theory (Aumann, 1987; Kreps, 1990a). The novelty of Axelrod's contribution was to make a computer experiment on the iterated Prisoner's Dilemma.[16] He invited the experts in social science and game theory to deliver programs each of which was in their opinion the winning strategy in a computer tournament between the different programs. The results of this first tournament (14 entries) were used as an invitation to a larger tournament in which 62 experts as well as computer hobbyists participated. The conditions of the two tournaments differed in minor respects but the main conditions of the competitions were:

1. The payoffs for the different alternatives are defined by table 5.2.
2. Each program will play with each other program 200 times, or approximately that number.[17]
3. Each program will also play 200 times against the program RANDOM (self-explaining) and against a copy of itself.
4. The program is allowed to have an arbitrary length.
5. The program has access to the history of the iterated game and may use this history in making a choice.

Table 5.2. The Prisoner's Dilemma.[18]

		Player B Cooperate (C)	Player B Defect (D)
Player A	Cooperate (C)	3, 3 Reward for mutual cooperation	0, 5 Sucker's payoff, and temptation to defect
Player A	Defect (D)	5, 0 Temptation to defect and sucker's payoff	1, 1 Punishment for mutual defection

The winner of both tournaments was the most simple program delivered, called TIT-FOR-TAT. Its very simple strategy (or decision-rule) is:

1. Cooperate on move 1.
2. Thereafter, do whatever the other player did in the previous move.

This strategy cannot win all plays. If a sufficient number of the other programs consistently defect, TIT-FOR-TAT will come out as a loser (since it will lose in the first round). However, it has the advantage of interacting reasonably well with a great number of other strategies. The reason for this robustness is that TIT-FOR-TAT has the following properties: niceness and clarity, provocability but forgiveness (Axelrod, 1984, ch. 2). The program ALL-D (always defect) is not nice and cannot exploit possibilities of cooperation. The program ALL-C (always cooperate) is nice but has no provocability and can easily be exploited by other programs. The program MASSIVE-RETALIATORY-STRIKE (cooperate until a defection is met, and then answer with a constant defection) is not forgiving and creates a situation which is locked into defecting behaviour. More complex programs like TWO-TITS-FOR-ONE-TAT may be able to avoid certain types of exploitation of TIT-FOR-TAT but are more difficult for other programs to understand; to cooperate with them other programs have to consider the two last moves in the game. The more complex the program, the less clear its structure is to other programs who want to cooperate with it.

In an evolutionary perspective, the computer tournament described so far includes 1) variety-creation by experts in game theory and 2) selection by means of the payoff matrix and the procedures of the tournament. To approach an evolutionary process, Axelrod included: 3) differential reproduction of programs (strategy applications) in an 'ecological tournament' (Axelrod, 1984, 48-52, 203-205, 217). We may think in terms of a series of tournaments. The programs which have done relatively well in one tournament will be represented by a larger number of copies (a larger 'clone') in the next tournament, while less successful programs will be represented by fewer copies, and they may eventually become extinct. The number of copies of a program will not necessarily show a steady upward or downward movement. Some programs (like ALL-D) did well in the early tournaments because easily exploitable programs were available (like ALL-C). When the exploitable programs become less numerous, then the exploiters will also decrease in number. Programs whose success is due to their ability to cooperate with others as well as themselves (like TIT-FOR-TAT) are not vulnerable to such a mechanism. However, as TIT-FOR-TAT become more numerous, ALL-D may have another chance (if it has not become extinct). Therefore, we cannot say that TIT-FOR-TAT is an 'evolutionary stable strategy' in the sense of Smith (1982, 204), i.e., a strategy which cannot be 'invaded' by a single copy of an alternative strategy if held by all other members of the population. Unfortunately, ALL-D is such an evolutionary stable strategy. If a single TIT-FOR-TAT is ever to be faced with a bunch of ALL-D's, then the nice strategy is bound to become extinct.

The experiment with iterated tournaments was called 'ecological' because it concerned the interaction and differential reproductive success of a fixed set of programs. In an evolutionary perspective this is a step backwards *vis-à-vis* Axelrod's first two computer tournaments where game theorists provided them a new variety of programs for selection among. However, Axelrod (1987) has explored the possibility of letting the computer do the job of the game theorists: to create new programs which may be fed into the evolutionary process. A method of doing so was outlined in section 5.1.1 (Holland, 1975; Davis, 1987). In this technique of genetic algorithms strategies represent a kind of 'chromosomes' or 'genotypes' with a dual purpose: First, the 'genotypes' may be interpreted in order to decide the behaviour of the program in a given environment (e.g., an iterated game). Second, the 'genotypes' may be manipulated in a systematic way when copies of the programs are made for the next computer tournament. We no longer have a simple 'clone' of program copies. Instead some of the offspring of a given program have a different 'genotype' and will show a modified behaviour in comparison with their parent program.

The genetic algorithm is clearly inspired by evolutionary biology (genetics) but it was generalised by Holland to become a general problem-solving technique. In relation to Axelrod's work and to evolutionary economics it may be considered as a theoretical tool which in certain situations helps us to introduce new variety into the evolutionary process. At least in relation to many types of strategy for playing the iterated Prisoner's Dilemma, it is not difficult to define the 'genotype' as well as ways of modifying it (see section 5.4.3). As a result we may start with a population of programs whose strategies are created at random. We do not know whether, e.g., ALL-P, ALL-D, TIT-FOR-TAT, TWO-TITS-FOR-ONE-TAT, etc. are present at the beginning. As the generations come and go, some characteristics of the successful programs may begin to appear. According to Axelrod (1987, 36-38) the new simulation results[19] are not incompatible with his earlier results about TIT-FOR-TAT or, more generally, about the kind of more or less cooperative strategy which is likely to succeed in the long run. However, a closer look at his short presentation reveals that significant differences have emerged. The full blossoming cooperative behaviour cannot be taken for granted, especially not in relatively stable situations where the strategies of different individuals in a certain generation are not allowed to mix in the creation of offspring.

The following presentation takes its point of departure in this analysis. No attempts will be made to draw upon the huge literature of potential interest to the evolution of strategies of transaction.[20] The tasks are much more modest: 1) to demonstrate the evolutionary-economic relevance of Axelrod's work through a reinterpretation and an extension of it; 2) to introduce important tools of evolutionary-economic analysis.

5.2.2. A fable of the difficult evolution of cooperation

To emphasise its economic relevance we may think of the Prisoner's Dilemma as a highly generalised version of what we may call the Trader's Dilemma when facing uncertainty about the quality of the products delivered. This problem has been discussed from Gresham to Akerlof (1970) and beyond. Actually, one of the inventors of the game had been experimenting with simple exchange games (used cars for money, see Flood, 1958), and the original creation of the Prisoner's Dilemma game may perhaps be seen as a simplification and generalisation of such exchanges rather than as a caricature of American criminal justice which tries to lure one gangster to desert his criminal partner (Hardin, 1990, 364).

Let us (in relation to the work of Axelrod and others) explore an evolutionary story or, better, a fable which may be developed from this reinterpretation of the game. By calling it a 'fable', we emphasise the highly stylised character of the kinds of problems that we deal with. It is a kind of highly abstract story-telling with a grain of truth, but not even a highly abstract version of *A Theory of Economic History* (to use the title of a too little acknowledged book: Hicks, 1969). Just as good fairy tales are often about a small kingdom, so is this one: the Kingdom of Mercantia. Furthermore, it is as usual the Queen who defines the rules of the game. In this way she has the same role as the authorities in the Prisoner's Dilemma story. In another sense her role is exactly the opposite.

While the goal of the criminal authorities was to create defecting behaviour (with respect to the norms of the criminals), the strategy ALL-D is the primeval state of the economy in our fable about Mercantia as well as in Axelrod's analysis. It is a kind of economic version of the Hobbesian *bellum omnium contra omnes*, and the problem for the Queen is to change this state of nature which, among other things, decreases the ability of the merchants to pay taxes to their government. Therefore, the Queen discuss with her advisers which kind of reform is most appropriate for changing the situation. One of her advisers summarises an apparently similar debate in the more wealthy and stable Kingdom of Entropia. Here the King's conclusion was:

We can never expect economic life in Entropia to be easily predictable or perfectly stable. Inevitably, there will be many uncertainties. ... But if we are willing to breed or adopt innovations at the appropriate times, and not to cling too tightly to our old ways, then we may be assured of reaching a new balanced path of economic development! (Batten, 1987, 82 f.)

The Queen answers that the situation in her Kingdom of Mercantia is much too predictable. To emphasise her point she refers to a report about the state of the economy:

Even the crude monetary system of the country has broken down and instead barter trade prevails. All the n traders of the country love to cheat by delivering commodities of low quality which their counterparts cannot check at the moment of exchange. However, all the traders hate to find out afterwards that they have filled the role of a cheatable 'sucker'. Since they are neutral to a double deceit, they always exchange low quality commodities which give equal but low payoffs.[21] By studying the barter trade at the marketplace of Mercantia, the report-makers have found out that this trade can be divided into unit transactions which in terms of utility can be described as in table 5.2. The pure utility of a defective commodity is 2 while it is only 1 if the trader finds out that he has been one-sidedly cheated. Similarly, the pure utility of a perfect commodity is 6 while it is 7 if the trader knows that he has succeeded in a one-sided cheat. In this situation both traders have a dominant strategy, that is, a strategy that is the best reply to all the moves of the trading partner, and the strategic (Nash) equilibrium of the game is the mutual delivery of defective commodities. The dilemma for a trader is that he would be better off with mutual delivery of perfect commodities but as a rational trader he will chose the defective delivery, fully knowing that the partner will do the same.

Table 5.3. The Trader's Dilemma.[22]

		Trader B	
		Perfect (P)	Defective (D)
Trader A	Perfect (P)	6, 6	1, 7
	Defective (D)	7, 1	2, 2

After confirming these facts, the Queen proceeds to discuss different possibilities of changing the rules of the game. According to Hobbes, she is an absolute monarch who by necessity must control the antagonisms of individual interest. However, the liberal traditions of her country as well as the research of Axelrod suggest that the problem lies in the lack of repetition of the bilateral exchanges between the different traders. The merchants may be able to provide the necessary control of each other and thus protect trade contracts with very little intervention from the Queen. The goal is, of course, to triple the utility of the traders (see table 5.3). The macroeconomic outcomes cannot be predicted with much accuracy; but a substantial expansion of the tax base will undoubtedly be the result. Appeals to decency and honesty among the traders are not an effective kind of intervention; such measures have already failed. The attempt to eliminate the Trader's Dilemma by introducing a system of quality control has only led to still new ways of presenting defective commodities as if they were perfect. The conclusion of the Queen is that it is high time to redesign the rules of the game. The goal is to create an evolutionary process which changes the strategies of an iterated Trader's Game. The structure of the reformed system is summed up by the Queen in figure 5.1 (see next page).

In her presentation of figure 5.1, the Queen emphasises the following points. First, during each month, each of the traders must transact (approximately[23]) a fixed number of times with a fixed number of other traders. Each trader is supposed to make his decision about his behaviour in the next transaction on the basis of a strategy which is fixed for one month of trading. The strategy is supposed to make use of information from previous transactions (arrow 1 of figure 5.1).

Second, the trader must inform an officer (a notary) about the quality of the commodity received in each transaction after it has been revealed whether it is perfect or defective. The notary makes a note of this information in her records for the informing trader (arrow 2). Each trader has access to his own record and the notary thus helps him keep track of the behaviour revealed by his transactions.

158   *Evolutionary Economics*

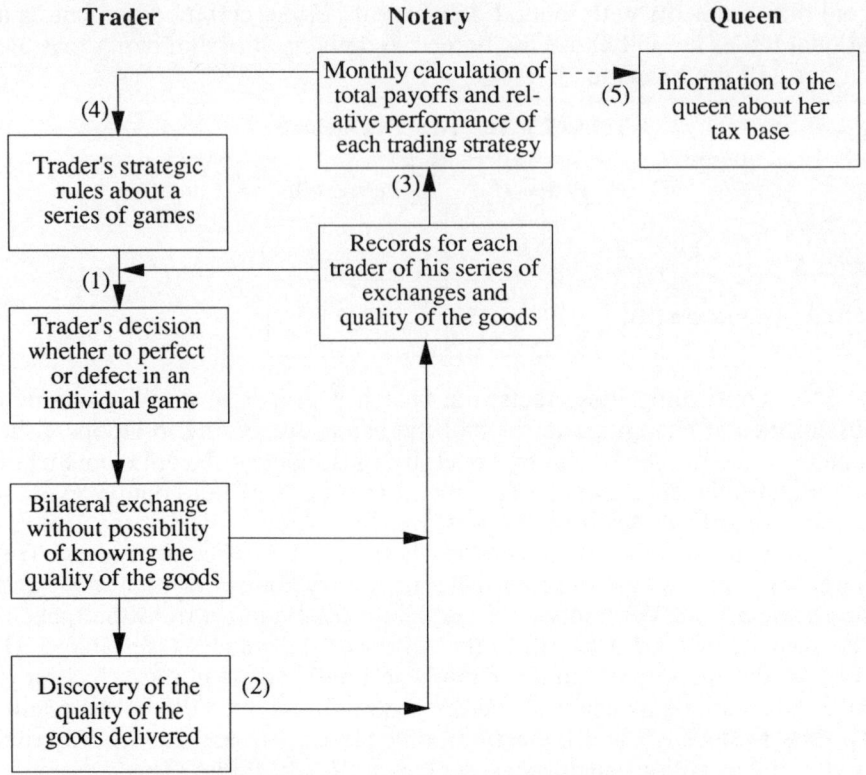

Figure 5.1. The set-up and part of the computational structure of the iterated Trader's Game after the Queen's reform.

Third, it is emphasised by the Queen that at the end of each month, the notary sums up the payoffs of all the transactions of all the traders (aggregation and belief in cardinal utility are wide-spread in the Kingdom of Mercantia); at the same time the notary calculates the relative performance of each strategy (arrow 3).

Fourth, the notary assigns each trader a 'competitiveness index' (the ratio between the trader's payoffs and the average payoffs) (arrow 4). Based on this information which reveals his relative performance, each trader is supposed to rethink his strategy. A low performance will increase his propensity to search for another strategy, either by innovative work or by attempts to get access to information about other traders' strategies as well as about their performance indices. Thus, high performance strategies become adopted by more traders while low performance strategies are gradually abandoned.

Fifth, the Queen is given an index which shows to which degree the Trader's Dilemma has been overcome. From another perspective this index describes the presumed willingness and/or ability of the traders to pay extra taxes.[24] Finally, it is emphasised that this procedure is to be repeated month after month, year after year, decade after decade.

After having presented her reform, the Queen asks for a last-minute analysis of the probable outcomes of the reform. Some of the experts emphasise that information becomes a free good to the traders, and that this may induce them to engage in very complex behaviour; the notary is an artificial construct just as the Walrasian auctioneer. Other experts emphasise uncertainty and ignorance in the study of basic social change. Less cautious experts prepare the following note (sections 5.3–5.4) which may at least give a framework for understanding the functioning of the reform as well as some forewarnings about what might happen to the reform.

5.3. Iterated games during a trading period with fixed rules

To understand the results of the Queen's reform we have to consider several characteristics of the game which will also provide us with several notions which are used to characterise evolutionary processes. Let us for simplicity assume that each month, each trader performs a sequence of (approximately) 100 unit transactions with each of the other 99 traders. This means that each trader makes a deal with each of the other traders a little more than three times a day. We may think in terms of trading rounds which include one exchange between all pairs of traders. A few hours after such a trading round, the qualities of the received commodities have been revealed and recorded, and a new trading round can begin.

5.3.1. The possibility of retrospection

In the following we will take the viewpoint of trader A who has to consider his exchanges with 99 other traders. In the initial exposition it is, however, sufficient to consider his exchanges with trader B. The decision-problem for trader A is whether to deliver a perfect or defective commodity to trader B in the $i+1$'th round of the trading game. Before the Queen's reform, trading was so sporadic and the recording of its results so unsystematic that each transaction was done without any reference to the historical records. This has now been changed since the basic idea of the Queen's reform is that it gives each trader the possibility of analysing the records of his exchanges with specified trading partners. The question now is what help in decision-making trader A can obtain from the records of his exchanges with B.

The quality of the commodities exchanged between A and B can be recorded as an array of pairs. Initially all entries are denoted by an 'empty' marker: '–'. When information is obtained, it is recorded in the next empty pair in the following way: Each pair starts with information about A's delivery to B in the i'th round of transactions, $T_{A|Bi}$. This variable is either "P" for a perfect commodity or "D" for a defective commodity. The second component of the pair for the i'th round is, of course, B's delivery to A, $T_{B|Ai}$. The whole array of information about the commodities exchanged from the first to the i'th round of transactions between A and B, T_{ABi}, may now be defined.

However, before we make the final decision about the specification of T_{ABi}, it should be noted that it is mainly the informations from the last few rounds which are of potential relevance to A's decision-making. When A makes the decision for the $i+1$'th round he may need information about the i'th round, the $i-1$'th round, and possibly a few more. The rest of the array will be ignored. Because we read text-strings from left-to-right, it is convenient to put information about the first pair of commodities to the right of the array and then to fill it out from right-to-left. Another practical point is that trader A will often chose to ignore his own deliveries and only consider the quality of B's deliveries. For this reason we also need an array, $T_{B|Ai}$, which simply is a list of B's deliveries to A from the first round to the i'th round. The busy notary wants a simple and distinct notation for this extra list and for this reason she uses the alphabet of digital computers: "1" for perfect and "0" for defective (this implementation will be of use later). Formally, we define our two arrays as:

$$T_{ABi} \equiv \left((T_{A|Bn}, T_{B|An}), \ldots, (T_{A|Bi}, T_{B|Ai}), \ldots, (T_{A|B1}, T_{B|A1})\right)$$
where
if $j \leq i$ then $T_{A|Bj}, T_{B|Aj} \in \{"P","D"\}$;
if $j > i$ then $T_{A|Bj}, T_{B|Aj} \in \{"-"\}$; [25] (5.1)

$$T_{B|Ai} \equiv \left(T_{B|An}, \ldots, T_{B|Ai}, \ldots, T_{B|A1}\right)$$
where
if $j \leq i$ then $T_{B|Aj} \in \{"1","0"\}$;
if $j > i$ then $T_{B|Aj} \in \{"-"\}$; (5.2)

We may now return to A's decision-making process concerning his exchange with B in $i+1$'th round of the trading game of a particular month. There are records for earlier months about the interaction between A and B. To simplify we will, however, assume that A decides not to take these records into account. The reason is that B may have changed his strategy at the end of last month. Any prejudice about B's behaviour may spoil the game.[26] Assume that the i'th transaction

($1 \leq i \leq n$) has been made and evaluated and the $i+1$'th transaction is to be decided upon. For this decision-making problem, A has information about the transaction rounds $i, i-1, i-2, i-3, \ldots$ For example, the records may show the following pattern: (P,P),(P,D),(D,P),(P,P),..., or, with respect to the deliveries of B to A: 1011... In words, B delivered perfect commodities in three of the four last exchange rounds. Only in the second to last ($i-1$'th) round, did he deliver a defective commodity. When we consider both trading partners we have the following story about the trading game: First ($i-3$), both A and B delivered perfect commodities, then ($i-2$), A defected while B had an unchanged behaviour, then ($i-1$), B defected while A returned to a delivery of a perfect commodity, finally (i), both delivered perfect commodities.

This information may be used in determining A's next move but first he has to interpret the information. He himself knows best whether he initially ($i-2$) defected by caprice or to explore B's behaviour. The interesting thing is B's reaction. One hypothesis is that there is a causal relation between A's defection $\left(T_{A|B,i-2} = D\right)$ and B's subsequent defection $\left(T_{B|A,i-1} = D\right)$. According to this interpretation, B punishes A for his defection. Furthermore, the relationship between $T_{A|B,i-2} = P$ and $T_{B|A,i-1} = P$ may suggest the hypothesis that B is 'forgiving'. If these hypotheses are correct, A will maximise his payoffs by delivering perfect goods in the rest of the game (provided that he is not in an 'end game' situation). However, A can never be sure about his hypotheses since B's moves may reflect earlier parts of the game and/or one or more randomly chosen moves (like A's supposed testing of B's reaction). He may try out his hypotheses but he has to reconsider the situation if his predictions are falsified by some future move by B. Thus, it is clear that this kind of retrospection is often quite difficult, while one glance will be sufficient to decide the typical play prior to reform, even after 99 transactions: (D,D),(D,D),(D,D),...,(D,D).

5.3.2. Strategies for the iterated Trader's Dilemma

Each trader is involved in so many transactions each month (say, 100 transactions with each of 99 other traders) that there are good reasons to believe that he will rely on procedurally determined behaviour and that he will change his procedure at most once a month when he is provided with information about his relative performance (see figure 5.1). Actually, the above recorded attempts by A to make guesses about the behaviour of B were based on the assumption that B was following a fixed and simple strategy. We have already met such strategies in the presentation of Axelrod's studies: ALL-P, ALL-D, TIT-FOR-TAT and TWO-TITS-FOR-ONE-TAT. The problem is now how to relate these strategies to the information available to trader A, to consider alternative strategies and to develop a general notation for strategies which makes possible their automatic interpretation.

The first task is simple. The simple-minded strategies are ALL-P and ALL-D; if following one of them, trader A has no need to look at the records since he already knows what to do. Such was the situation before the reform. More complex strategies need to consider the records. But taken individually, they are not difficult to handle. We may, e.g., define decision-functions for TIT-FOR-TAT and TWO-TITS-FOR-ONE-TAT. In both cases the input to the function is A's records up to round i, T_{ABi}, and the output from the function is A's move in round $i+1$, i.e., $T_{A|B,i+1}$. In the case of TIT-FOR-TAT only B's last move, $T_{B|Ai}$, is considered and A simply copies this move; the exception is the first move where A's move is "P".[27] In the case of TWO-TITS-FOR-ONE-TAT the strategy takes into account the two last moves. The function only delivers the output "P" if B's two last moves have this value. This gives the following definitions of TIT-FOR-TAT (5.3) and TWO-TITS-FOR-ONE-TAT (5.4). For reasons to be explained below they are named $\overline{PD}$ and $\overline{PDDD}$:

$$\left. \begin{array}{c} \overline{PD}(T_{ABi}) \equiv \overline{PD}\left(\text{input}:T_{ABi}, \text{output}:T_{A|B,i+1}\right) \\ \underline{\text{if}}\ T_{B|Ai} = \text{"--"}\ \underline{\text{then}}\ T_{A|B,i+1} := \text{"P"} \\ \underline{\text{else}}\ T_{A|B,i+1} := T_{B|Ai}; \end{array} \right\} \qquad (5.3)$$

$$\left. \begin{array}{c} \overline{PDDD}(T_{ABi}) \equiv \overline{PDDD}\left(\text{input}:T_{ABi}, \text{output}:T_{A|B,i+1}\right) \\ \underline{\text{if}}\ T_{B|Ai} = \text{"P"}\ \underline{\text{and}}\ T_{B|A,i-1} = \text{"P"} \\ \underline{\text{then}}\ T_{A|B,i+1} := \text{"P"} \\ \underline{\text{else}}\ T_{A|B,i+1} := \text{"D"}; \end{array} \right\}^{28} \qquad (5.4)$$

where $\overline{PD}(T_{ABi})$ and $\overline{PDDD}(T_{ABi})$ are the formats to be used when the functions TIT-FOR-TAT and TWO-TITS-FOR-ONE-TAT are applied; the similar expressions on the right side specify the input to as well as the output from the functions; the 'body' of the functions specifies how the output value is generated from the input; in (5.3) it is first checked whether the input is empty: if it is, then the output is "P", else output is a copy of B's last move; in (5.4) output is only "P" if the two last moves of B are "P"; the 'opening game' is not specified.

The specifications of these simple strategies have used a notation of strategies which has not yet been presented. The reason for such a notation is that an extreme variety of strategies is available to trader A; we are only able to keep track of these strategies (and to generate new ones through the genetic algorithm), if we have a fully specified notation.

The evolution of strategies of transaction 163

This notation is illustrated in table 5.4 which describes strategies of A which take into account exactly the last move, the two last moves, and the three last moves of B ('opening game' variants are ignored).

Table 5.4. Notation and examples of pure (non-random) strategies taking into account the last (a), the last two (b), and the last three (c) moves of the trading partner.

(a) 2-bit strategies

Trader A's strategies (on A's behaviour in $i+1$)	Trader B's behaviour in round i	
	P	D
ALL-D ($\overline{DD}$)	D	D
ALL-P ($\overline{PP}$)	P	P
TIT-FOR-TAT ($\overline{PD}$)	P	D
ANTI-TIT-FOR-TAT ($\overline{DP}$)	D	P

(b) 4-bit strategies

Trader A's strategies (on A's behaviour in $i+1$)	Trader B's behaviour in round i and $i-1$			
	PP	PD	DP	DD
TIT-FOR-TWO-TATS ($\overline{PPPD}$)	P	P	P	D
TWO-TITS-FOR-ONE-TAT ($\overline{PDDD}$)	P	D	D	D
Unnamed ($\overline{PDPD}$)	P	D	P	D
(+13 more strategies)				

(c) 8-bit strategies

Trader A's strategies (on A's behaviour in $i+1$)	Trader B's behaviour in i, $i-1$ and $i-2$							
	PPP	PPD	PDP	PDD	DPP	DPD	DDP	DDD
$\overline{PPPPPPDD}$	P	P	P	P	P	P	D	D
$\overline{PPDDPDDD}$	P	P	D	D	P	D	D	D
(+254 more strategies)								

In table 5.4.a all four possible strategies which respond to B's last move are shown. For example, we see that ALL-D can be described as $\overline{DD}$ since the answer to B's moves "P" and "D" are both "D". Similarly, TIT-FOR-TAT can be described as $\overline{PD}$ since the answer to B's move "P" is "P" and the answer to B's move "D" is "D". In table 5.4.b we see 3 of the 16 strategies which take into account exactly the last two moves of B. Such strategies are specified by 4 bits of information. For instance, TWO-TITS-FOR-ONE-TAT can be described as $\overline{PDDD}$ since the answer to B's moves "P" in the last and "P" in the next last round is "P" while the answer to all other histories of moves is "D". In table 5.4.c are provided a couple of examples of the 256 strategies which take into account exactly the three last moves of B. Such strategies are specified by 8 bits of information.

Table 5.4 demonstrates a notation which unambiguously describes an enormous number of strategies for the iterated Trader's Dilemma. The main purposes of this notation are 1) to help study the complexity of trader A's search for strategies (section 5.3.3), and 2) to formulate strategies in a form which is producible and modifiable by the genetic algorithm (section 5.4.3). The notation proposed here is only one of several possibilities., of course, alternative notations.

5.3.3. Complexity and bounded rationality

Until now, we have studied a type of strategy for trader A which only takes into account a certain number of previous moves of trader B. Even in this simple case, the message is clear. The more information A takes into account in his decision-making, the larger is the space of available strategies. The number of possible strategies will rapidly become so large that any global knowledge is impossible. Consider the sequence of strategy types started in table 5.4: If A takes into account 1 move of B (table 5.4.a), then there are 2 different histories of B's behaviour; A's response takes the form of 4 possible strategies which are describable by a 2-bit string $(\overline{PP}, \overline{PD}, \overline{DP}, \overline{DD};$ or: $\overline{11}, \overline{10}, \overline{01}, \overline{00})$. When considering B's 2 last moves (table 5.4.b), there are 4 histories and A has 16 possibilities which are describable by a 4-bit string. B's 3 last moves (table 5.4.c) have 8 different histories; A has 256 possible strategies to be described by an 8-bit string (like, e.g., $\overline{PPPPPPDD}$). B's 4 last moves create 16 possible histories, 65536 possible A-strategies and a 16-bit notation. The 5 last moves give 32 histories and more than 4 thousand million strategies, etc. In this way we may proceed until we come to a strategy of A's which is able to cope with all the information available, even at the very end of the game. In this case, A must be able to consider (approximately) 99 moves of B. The number of possible strategies is astronomical.

We may describe all these strategies in a general way by means of a modified version of the so-called classifier systems used in Artificial Intelligence research (see section 5.1.1). Basically, we are dealing with a variant of a rule-based production system (like the ones used in expert systems). They may be considered as databases of rules of the form: "if X then Y;" where the condition X is a history of B's moves while the response Y is A's behaviour in the next move. When the rule-based production system receives a history (a message), a search is made to find the appropriate rule which then gives the answer. The history or message is formulated according to definition (5.2). A general strategy must know which bits of information it is to consider and which to ignore. The former may be denoted by "P" and "D" or "1" and "0". The bits to be ignored by the decision-making process are denoted by "#". In these terms, we may give the following general form of a definition of a strategy which takes into account the last λ moves of trader B:

$$\overline{S_1 S_2 ... S_{2^\lambda}}\left(T_{B|A\tau}\right)$$

$$\equiv \overline{S_1 S_2 ... S_j ... S_{2^\lambda}}\left(\underline{\text{input: }} T_{B|A\tau}, \underline{\text{output:}} T_{A|B,\tau+1}\right)$$

where

$S_j \in \{"P","D","\#"\};$

$1 \leq \lambda < 100;$

$T_{B|A\tau j} \in \{"P","D"\} \underline{\text{if }} j \leq \tau;$

$T_{B|A\tau j} \in \{"-"\} \underline{\text{if }} \tau < j < 100;$

[29] (5.5)

where S_1 is the first symbol in A's strategy-definition; λ is the number of B's moves which is at most taken into consideration by A's decision; S_{2^λ} is the last symbol of the strategy-definition; if, e.g., λ is 4, then number of symbols is $2^4 = 16$; τ is an indicator of the last of the rounds recorded by the information of B's previous moves; of course, the string of input information is empty for moves larger than τ.

With this notation we are able to describe all possible strategies which produce A's decision in a deterministic manner by studying B's moves. But this is only one type of strategy. One problem is that we have not enumerated the increasing number of 'opening game' variants. More importantly, we have been dealing with a type of strategy which considers only half of the available information, namely the list of moves from the trading partner B. No attention has been given to the interdependence between A's and B's moves. This interdependence was only verbally presented in section 5.3.1 and available as a strategic possibility through the information defined in (5.1). The type of strategy considered here could as well have been defined in terms of the list of B's moves defined in (5.2). Actually, we will do so in most of this chapter. At present it may be helpful to demonstrate the enormous number of strategies available which include both A's and B's moves.

An example may suffice: the strategies which use information from exactly the last three moves of A as well as of B (the case used by Axelrod, 1987, see section 5.2). Since there are four possible outcomes of each transaction, we have 64 different possible histories of the three last transactions. Each history is described in terms of an 8-bit array of "P"s and "D"s. The interpretation of one such history was discussed in section 5.3.1 where A made an experimental default and considered B's reaction: the case "PP,PD,DP,PP" or "11,10,01,11". This case was simple in itself, even if it included some judgements. But if A is to be able to respond automatically to all the different histories, he must specify a 64-bit strategy, e.g., a version of the ALL-D strategy: $\overline{DDDD...D}$. Furthermore, this specification of the game which takes into account the moves of exactly three rounds of transactions between A and B, necessitates that each participant makes three preliminary and independent moves before the real game can start. To specify these possibilities 6 extra bits are necessary. A strategy must thus be formulated in terms of 70 "P"s and "D"s where each responds to a particular history.

All this is simple. But it means that even decision-making which considers the results of the three last transactions must chose from 2^{70} different strategies. To emphasise the impossibility of exploring this huge search space (which is a tiny bit of the whole search space), Axelrod (1987, 34) remarks: 'If a computer had examined these strategies at the rate of 100 per second since the beginning of the universe, less than one percent would have been checked by now.' The conclusion is, of course, that the Trader's Game is characterised by boundedly rational behaviour. One of the heuristics is to delimit the analysis to B's moves. Another is to think in terms of Axelrod's four characteristics of the kind of strategy which succeeded in his computer tournaments: such a strategy might be nice, provocable, forgiving and clear to the trading partners. But we really do not know. Computer simulations of the evolution of strategies of transaction may give some rough answers.

5.3.4. The conditional values of Strategies

It is now time to consider the payoffs of different strategies by means of the payoff matrix of the Trader's Dilemma game (table 5.3). This information is provided by the notary of our fable (figure 5.1) but in principle it can also be calculated by trader A and the other traders. Assuming that only the moves of the trading partner are taken into account, we may return to our simple system of strategies. Furthermore, we repeat the assumption that the strategies are fixed during a whole iterated game (supergame) between a pair of trading partners. Thus, we may have $\overline{DD}$ vs. $\overline{DD}$, $\overline{DD}$ vs. $\overline{PD}$, $\overline{DD}$ vs. $\overline{PP}$, etc. For each type of iterated game it is easy to calculate the payoffs for the traders (by use of table 5.3). In the situation before the reform of the trading system, we always saw $\overline{DD}$ vs. $\overline{DD}$, and the cumulative result of a hundred transactions would be 200 utility units for both partners, or $V_{100}(\overline{DD}|\overline{DD}) = 200$. However, all other supergames in which $\overline{DD}$ participates have asymmetric payoffs with $\overline{DD}$ as a winner. It gains seven times as much as the 'naive' $\overline{PP}$, and a little more than $\overline{PD}$ (TIT-FOR-TAT) because the latter starts by exchanging a perfect commodity against a defective one $\left(V_n(\overline{DD}|\overline{PD}) > V_n(\overline{PD}|\overline{DD})\right)$. All other strategies have a more mixed performance. To illustrate this kind of result, table 5.5 shows the supergame payoffs $\left(V_{100}(\overline{X}|\overline{Y})\right)$ for the four strategies which only take into account the behaviour of the trading partner in the last transaction.

Table 5.5. The cumulated payoffs of 100 transactions between two traders, according to the strategies chosen.[30]

		Trader B			
		$\overline{PP}$	$\overline{PP}$	$\overline{DP}$	$\overline{DD}$
Trader A	$\overline{PP}$	600, 600	600, 600	100, 700	100, 700
	$\overline{PD}$	600, 600	600, 600	400, 400	199, 205
	$\overline{DP}$	700, 100	400, 400	350, 350	101, 696
	$\overline{DD}$	700, 100	205, 199	695, 101	200, 200

In table 5.5 we, for example, see that $\overline{PD}$ does not beat any other strategy but collaborates perfectly with itself as well as $\overline{PP}$ while it trades fairly well with $\overline{DP}$ and is only slightly beaten by $\overline{DD}$. Similar supergame payoffs can be calculated for the more complex strategies like $\overline{PPPD}$, $\overline{PDDD}$ $\overline{PPPPPPDD}$, etc.

However, it must be emphasised that these calculations presuppose perfect information about the quality of the commodities and different payoffs for some of the strategies may be observed if we introduce a little 'noise' into the game. As an example we have the cases where trader A by mistake delivers a defective commodity or where trader B by mistake registers a defective delivery. Such noise will marginally change the payoffs of table 5.5. But imagine the MASSIVE-RETALIATORY-STRIKE strategy which delivers defective commodities for the rest of the game if the partner ever delivers a defective commodity. In a noiseless supergame between a $\overline{PP}$-strategist and a MASSIVE-RETALIATORY-STRIKE-strategist the payoffs will be 600 while a little noise may lead to a number between 600 and 100 $(100 < V_{100}("\text{massive}-D"|\overline{PP}+"\text{noise}") \leq 600)$[31].

5.3.5. The overall value of a strategy for a month's trading

Trader A is not only following a fixed strategy in the supergame with trader B but also in the supergames with all the other traders. It is not the value of an individual supergame but the value of all the transactions during a month which is of interest. Let us call this value $V_t("\text{A's strategy}_t"|\bullet)$ where '$\bullet$' denotes all the other traders and t refers to the tth month after the start of the Queen's reform. The trading in each month may be considered as a 'tournament' between representatives of different strategies. It is clearly of importance whether most of the traders follow, e.g., $\overline{PP}$ or $\overline{DD}$. In other words, the value of the whole tournament for A depends on the strategy mix among the other traders. This dependence may not be easy to cope with. But the notary of our fable (see figure 5.1) ensures that the job is done: at the end of each month she makes the necessary calculations and gives information to the traders in a way which probably demonstrates the incentives for any change of strategy. In the real commercial world we have instead profitability indicators, etc. but such factors would complicate our story. Let us instead summarise the work done at the end of each month by our imaginary notary.

From the viewpoint of trader A, the notary's records cover his series of supergames each month, one with each of the $p-1$ other traders. Presupposing that each trader has applied the same (possibly extremely complex) strategy in all of his transactions during the previous month, the index is an evaluation of A's strategy in a fixed environment. The environment of the strategy chosen by trader A consists of the strategies chosen by the rest of the traders.[32] Instead of thinking of the other traders, we should just observe that the environment consists of strategies with different numbers of applications. To develop this perspective, we should 1) think in terms of relatively simple strategies and 2) imagine that the notary calculates the 'competitiveness index'. This can be done in many ways; as an example we may consider the following, rather boring, procedure which demonstrates the computer-like routines of the notary. After each period t, she applies the procedure for all the p members of the population of traders:[33]

The first task for the notary is to find out which strategy is followed by each trader. The best thing to do would be to ask him, but she can supplement that information by inspecting her records. For each of the pure strategies $(\bar{1},...,\bar{j},...,\bar{m}_t)$ followed during the month, she may determine how many traders have applied it $(N_{\bar{j}t})$.[34] The next task is for each individual strategy $\bar{s}$ to calculate (by means of tables like 5.5) the aggregate utility of each of the traders who have followed this strategy,

$$V_t(\bar{s}|\bullet) := (N_{\bar{s}t}-1)V_{100}(\bar{s}|\bar{s}) + \sum_{\bar{j}=1, \bar{j}\neq\bar{s}}^{\bar{m}_t} N_{\bar{j}t}V_{100}(\bar{s}|\bar{j});[35] \quad (5.6)$$

where $V_{100}(\bar{s}|\bar{j})$ is the value of 100 single games/transactions between a trader following strategy $\bar{s}$ and a trader following strategy $\bar{j}$, and $N_{\bar{j}t}$ is the number of supergames of this type that each trader has; within his own strategy group there is one supergame less since he does not play with himself.

Furthermore, the notary must calculate the average utility gained by the p traders during the period t,

$$V_t(\bullet|\bullet) := \frac{1}{p}\sum_{\bar{s}=1}^{\bar{m}_t} N_{\bar{s}t}V_t(\bar{s}|\bullet); \quad (5.7)$$

where the total utility obtained by an $\bar{s}$-strategist, $V_t(\bar{s}|\bullet)$, is multiplied by the number of $\bar{s}$-strategists, N_{it}, to give an aggregate for this group.

Then the notary calculates the total utility for all strategy groups and the average payoffs for a trader during period t. Finally, she calculates the 'competitiveness index' which a trader obtains for following each strategy,

$$W_t(\bar{s}|\bullet) := \frac{V_t(\bar{s}|\bullet)}{V_t(\bullet|\bullet)};\qquad(5.8)$$

which informs a trader following strategy $\bar{s}$ about his strategy-determined performance in the environment of strategies existing during last period (month).

This 'competitiveness index' is only interesting in heterogeneous populations of traders. In the case that all 100 traders have chosen $\overline{DD}$, the notary could immediately give all their indices, $W_t(\overline{DD}|99\overline{DD}) = 1,0$. However, a single $\overline{DD}$-strategist in a $\overline{PD}$-population would be informed that he is doing badly since $W_t(\overline{DD}|99\overline{PD}) = 0,3$. This information may surprise our $\overline{DD}$-strategist since the notary gives no information about the frequency of the strategies and since $\overline{DD}$ is a (narrow) winner of all his games with the other traders (see table 5.5). However, some reflection will make it clear to him that practically all the other traders have somehow escaped the Trader's Dilemma in the part of their transactions in which $\overline{DD}$ is not involved. This thinking, however, concerns an *ex post* ranking of the strategies and there is no guarantee that it helps to 'pick the winner' of last month since the simultaneous choices of other traders may change the environment in a crucial way. Furthermore, we should note that the notary keeps her ranking of strategies to herself. She only gives to each trader his personal 'competitiveness index'.

5.4. The evolutionary change of strategies

We have now set the stage for an evolutionary change of the strategies applied to the Trader's Dilemma game. The calculations of the payoffs are not too difficult provided that all the traders follow a fixed strategy in all their transactions during a month. The background for this type of rule-based decision-making is the incredible number of strategies available for even simple forms of decision-making of the traders. In this situation it seems to be rational to let a chosen strategy 'work itself out'. When information about the total utility and relative performance is obtained it may, however, be time for a change. Since all the traders may shift their strategy, there is much uncertainty about the environment in which the new strategy will function. Only a listing of the strategies chosen by all the other traders for the next month would allow a trader to choose the best strategy. Instead he has to make do with knowledge of the relative success or failure of his strategy from the last month, some scattered and imprecise knowledge about the previous strategies of the other traders and some ideas about the inertia of the system of strategies. But as long as the traders have not been locked into an evolutionary stable strategy (by all following $\overline{DD}$), it is rational to search for alternatives. For example, a follower of the altruistic $\overline{PP}$-strategy will do well in a population dominated by $\overline{PD}$-strategists (TIT-FOR-TAT), while $\overline{PP}$ fares badly in a population dominated by $\overline{DP}$-strategists (ANTI-TIT-FOR-TAT). The 'competitiveness index' given by the notary to each of the p traders at the end of each month is intended to depict this changing environment of the individual strategies. We now turn to an analysis of this evolutionary process.

5.4.1. The changing frequency of the different strategies

The basis of evolutionary analysis is a description of how the frequency of the application of the different strategies changes in the long run. If $N_{\bar{s}t}$ is the number of traders applying strategy $\bar{s}$ in period t, and p is the total number of traders,[36] then we may study the frequency $N_{\bar{s}t}/p$. Since the population of traders is constant in our case, the rate of change in the frequency, $w_{\bar{s}t}$, of strategy $\bar{s}$ in period t is simply

$$w_{\bar{s}t} := \frac{N_{\bar{s}t}/p_t}{N_{\bar{s},t-1}/p_{t-1}} = \frac{N_{\bar{s}t}}{N_{\bar{s},t-1}}; \qquad (5.9)$$

which corresponds to what in evolutionary studies is called 'observed fitness'. 'Evolution' may be said to have taken place if $w_{\bar{s}t} \neq 1$, for some $\bar{s}$ belonging to the set of strategies applied in period $t-1$. If the relative frequencies of the different strategies in an evolutionary process ever remain stable ($w_{\bar{s}t} = 1$, for all $\bar{s}$) for long periods, we may say that the evolutionary process has moved into a stasis (or long-term 'equilibrium').

In our fable the notary may amuse herself by following the changes in the frequencies of the different strategies but any publication of the results would spoil the game (especially because the traders would gain by giving false information to the notary). For all other persons in the economy, the 'observed fitness', $w_{\bar{s}t}$, cannot be exactly observed. Still it is in principle possible to measure this figure: it is a property of the real economic system. Even the researchers in the Kingdom of our fable will probably have to develop their analysis on the basis of relatively limited observations. However, the fable is quite suited for modelling and simulations which may be checked against the limited empirical observations.

The reason why we call $w_{\bar{s}t}$ the 'observed fitness' of strategy $\bar{s}$, is because we believe that environmental factors determine the major part of the change in the frequencies of application of strategies in the trader's world. But this is a hypothesis which has to be checked against the mechanisms generating change in the application of strategies. If we were to find out that all strategy change is due to random factors, the term 'observed fitness' would be a total misnomer. In any case, it is just a name for a well-defined measure (statement 5.9). The research task is to explain the observed facts. The whole set-up of the reformed Trader's Dilemma game is, however, designed to allow the environment, defined as the mix of strategies in month $t - 1$, to influence the frequencies of strategies in month t. But how and with what consequences? These are the questions that researchers hypothesise about. Their job is reflected by the notion of 'theoretical fitness', that is, the researcher's expectations about the changes in frequency of the application of a specific strategy. To emphasise this, we will now take information about the frequencies of different strategies during the first t months of the trading game as given and try to make hypotheses about the situation in the next month, $t + 1$. However, we will also use the hypotheses to make simulations of the development of the trading game.

As the game has been defined an increase of $N_{\bar{s}t}$ from period t to period $t + 1$ means that some of the traders who followed a strategy $\neq \bar{s}$ one month, have changed strategy and follow $\bar{s}$ during the next month. They may simply have made a mistake and this possibility is especially relevant for more complex strategies. This is the way the mutations take place in biological evolution, and in principle they cannot be neglected in the present case. We will, however, assume that traders make more conscious changes in strategies. Here we have two separate questions: Why does a trader want to change his strategy? Why does he choose a particular new strategy among the enormous number of possible ones?

Concerning the first question the 'competitiveness index' given to each of the traders at the end of each month gives us a clue. Take the case where a $\overline{DD}$-strategist sees that his 'competitiveness index' during some months or years falls from 1,0 to 0,3. This information may induce him to shift to another strategy and see whether this change will cause an increase in his index for the next month. As the present trading system is described, there are no mechanisms (e.g., relating to bankruptcy and the emergence of new traders) which strictly enforce such a change in strategy. However, we (and the Queen of our fable) may assume that the probability of a strategy change is positively correlated with the 'competitiveness index'. For example, we may formulate the simple hypothesis that the probability that a trader will uphold his strategy is proportionate to his 'competitiveness index'. Such an assumption is clearly related to individual rationality since a relatively bad performance indicates that strategies exist which have performed better during the last month. Whether they will do so in the next month's index depends upon the speed of change in strategies. Because we assume that this is relatively low, we have an environment which normally changes slowly.[37] Arguments for this slowness may be related to the difficulties of changing strategy; for example, there may be many errors in the months after a new strategy is chosen and these errors can only be overcome through an extended learning process.

We now turn to the second question: Why is one particular new strategy chosen? Even if there are reasons for changing strategies which are performing badly, the problem is how to find a better one in the huge search space (see section 5.3.3). There are three ways of finding a new strategy: imitation, local innovation (through recombination and mutation of existing strategies) and innovation by random choice in the space of possibilities. The imitative meta-strategy presupposes some degree of information exchange between the traders. An analysis of the supergames with the other traders will often give relatively well-founded hypotheses about which strategies are chosen by trading partners. Even if the value and 'competitiveness index' of each alternative strategy is not known exactly, some informed guesses may be made which increase the chance that changing a strategy will lead to success rather than failure. The meta-strategy of local innovation may allow a relatively systematic learning process about a slowly changing environment: $\overline{P\underline{P}DDPDDD}$ used during one month may be changed to $\overline{P\underline{D}DDPDDD}$. If the time perspective of the traders is long enough (i.e., if we have sufficiently small discounting factors), then such experiments may pay in certain situations. The meta-strategy of globally oriented innovation (implying, e.g., a jump from $\overline{DP}$ to $\overline{PPDDPDDD}$), is much more dubious. Should all traders chose this meta-strategy a sufficient number of times, the result would not be evolution but chaos: most of the strategies chosen would be so complex that the traders would be led to believe that nearly all their partners were behaving totally at random or according to different kinds of 'crazy logic'.

5.4.2. The theoretical fitness of strategies

To account for the role of imitation in the long-term evolution of the applications of strategies in the game, we must make specific hypotheses about the character of imitation as well as (localised) innovation. This evolution is the result of individual strategy choices[38] based on experiences of period t (and earlier periods) as well as random factors. Thus, the number of traders who apply strategy $\bar{s}$ in period $t + 1$ may be determined by:

1. The number of traders who decide to remain loyal to it.
2. The number who abandon it in order to imitate another strategy.
3. The number who abandon it because of their innovation of a new strategy (at least new to themselves).
4. The number who have left other strategies because they have decided to imitate the traders who used this strategy in period t.
5. The number who happen to reinvent strategy $\bar{s}$ through their process of innovation.

It is clear that there is a basic difference between on the one hand the group including conservatives and imitators (1, 2, 4) and on the other hand the innovative group (3, 5). The latter will in the main be dealt with in section 5.4.3.

In order to understand the processes underlying loyalty to or imitation of a given strategy $\bar{s}$, we may formulate the hypothesis that these processes will mainly depend upon the inducement to change away from strategies with a low 'competitiveness index' as well as the inducement to change towards strategies with a high 'competitiveness index'. However, we may also use the information underlying the 'competitiveness index' to articulate our hypothesis closer to the traditions of the mathematical theory of biological selection (first presented by R. A. Fisher). The basic equation may be articulated as presented in statements (5.6) and (5.7). The equation determines the hypothetical change in the proportion of the population which applies a strategy based on the difference between the utility obtained by applying strategy $\bar{s}$ and the average utility obtained by a trader. Thus, the change in the proportion of the traders who apply strategy $\bar{s}$ is

$$\frac{N_{\bar{s},t+1}^{\text{hypo}}}{p} - \frac{N_{\bar{s}t}}{p} := a\frac{N_{\bar{s}t}}{p}\left[V_t(\bar{s}|\bullet) - V_t(\bullet|\bullet)\right]; \tag{5.10}$$

where the difference between the utility of $\bar{s}$ and the average utility (see (5.6)–(5.7)) is multiplied with the proportion of traders applying $\bar{s}$ in period t and by a constant a. Furthermore, we remember that

$$\sum_{\bar{s}=1}^{\overline{m}_t} \frac{N_{\bar{s}t}}{p} = \sum_{\bar{s}=1}^{\overline{m}_{t+1}} \frac{N_{\bar{s},t+1}}{p} = 1.$$

The result is the hypothetical change in the proportion of traders applying this strategy between period t and period $t + 1$. Since a is the net result of entries to and exits from the strategies, we may be more explicit about our hypotheses ($a = a^{\text{entry}} - a^{\text{exit}}$) but there is no room for this at present.[39] We may even want to be less specific about our hypotheses and choose the following formulation:

$$\frac{N_{\bar{s},t+1}^{\text{hypo}}}{N_{\bar{s}t}} := f(W_t(\bar{s}|\bullet)); \tag{5.11}$$

where $W_t(\bar{s}|\bullet)$ is the 'competitiveness index' of $\bar{s}$ defined in (5.8).

By means of these hypotheses we may construct models and simulate the long-term (more than one month) development of the frequencies of the different strategies. However, the kind of change we would be studying is not an evolutionary but an 'ecological' process where the application frequencies of a *given* set of strategies change over time. If one strategy should ever become extinct, it will never reappear again. We now turn to innovation, which changes this property of our simulations.

5.4.3. The modified genetic algorithm

Hypotheses about the process of innovation may be formulated in several ways. As mentioned earlier (section 5.3.3), we may think of it in analogy to biological variety-creation where the genetic code of a strategy is seen as a string of binary information ($\overline{\text{PPDDPDDD}}$ or $\overline{11001000}$). Thus a 'genetic' string may be changed according to algorithmically specified meta-strategies (i.e., strategies for changing strategies). The different ways of making changes may be considered as different innovation strategies. However, we should be aware that the use of the biological analogy is normally related to the genetics of sexual reproduction. This is the case in Holland's (1975; Holland et al., 1986) version of the genetic algorithm and even Axelrod (1987) takes this as his primary case. But in economic affairs this is quite confusing[40] and is not overcome in recent contributions (see the survey in Marks, 1992, 33-36). Instead, our basic model should be based on asexual (or haploid) reproduction, and exchange of 'genetic' materials between our traders should be argued from an explicitly social context.

In the context of a modified version of the genetic algorithm scheme, we may define four idealised strategies for innovation which may be used by the traders:

1. Mutation: A single bit of information in the strategy string is changed; either $\overline{P}$ is changed to $\overline{D}$ or $\overline{D}$ is changed to $\overline{P}$. We have already seen the example where $\overline{P\underline{P}\text{DDPDDD}}$ is changed to $\overline{P\underline{D}\text{DDPDDD}}$.
2. Recombination: The strategy string is split up and recombined. For example, $\overline{\text{PPDD}}\,\overline{\text{PDDD}}$ is changed to $\overline{\text{PDDD}}\,\overline{\text{PPDD}}$.[41]
3. Inversion: The sequence of bits between two randomly assigned points in the strategy string is inverted (read backwards). For example, $\overline{\text{PPDD}}\,\overline{\text{PDDD}}$ is changed to $\overline{\text{DDPP}}\,\overline{\text{PDDD}}$.
4. Doubling: A copy of the strategy string is made and concatenated with its twin. For example, $\overline{\text{PPDDPDDD}}$ is changed to $\overline{\text{PPDDPDDD}}\,\overline{\text{PPDDPDDD}}$.[42]

In this set of ways to innovation, the most basic ones are mutation and doubling. Without them, many possibilities could not be exploited. The doubling strategy may be interpreted in the following way: in the beginning of strategy-construction all but the last transaction is ignored (e.g., $\overline{\text{PD\#\#\#...\#}}$); then the number of moves taken into consideration may gradually increase (e.g., $\overline{\text{PD}\underline{\text{PD}}\text{\#...\#}}$); and finally complex versions are reached (like the example above). The effect of recombination and inversion is to increase the speed by which different possibilities in the search space are tested in trading practice. These possibilities are, of course, very interesting when genetic algorithms are used for optimisation.[43] However, in an evolutionary-economic context they appear more dubious. In particular, inversion seems to lack a relevant interpretation in the traders' practice and it will in the main be left out of consideration.

Different kinds of innovation take place while the traders are considering their strategies at the end of a trading round (month). If the traders are characterised by satisficing behaviour, we may expect the probability that they will innovate to be dependent on their past performance. To make a simulation of the evolution of the strategies of the iterated Trader's Dilemma, we need to specify the probability of an innovation. Let us assume that the probability that an innovation will take place at the end of period t is the outcome of two kinds of considerations. First, there is the 'competitiveness index' of the strategy followed in period t which shows whether a given strategy has done worse or better than average. Second, there is room for absolute improvements which are reflected by the ratio between the utility obtained by using a given strategy and the maximum utility which can be imagined (see table 5.3). Formally we have

$$\gamma_{\bar{s}t}^n := f^n\left(W_t(\bar{s}|\bullet), \frac{V_t(\bar{s}|\bullet)}{V^{\max}}\right); \qquad (5.12)$$

where $\gamma_{\bar{s}t}^n$ is the probability to innovate given the strategy $\bar{s}$ in period t; $V^{\max}$ is the highest imaginable utility; and the other variables are found by the notary's calculations (see statements (5.8) and (5.6)). The introduction of the maximum outcome of the game is partly motivated by the case where all are playing $\overline{DD}$ and all have $W_t(\overline{DD}|\bullet) = 1,0$. However, it must be obvious to all the traders that the maximal utility will only be gained under very special conditions and that one should not put too much emphasis on the distance to it.

If an innovation is 'drawn', the outcome is determined by decision parameters and stochastic functions (see chapter 6). The parameters concern the propensities of a trader to chose mutation, recombination, inversion, and doubling. The kind of innovation that will be made can be determined probabilistically. In that case, the decision about which bit or bits to change will be determined by stochastic algorithms. Finally, the trader has a new strategy for the next period. As expected under the conditions of satisficing behaviour, this new strategy has taken its departure in the present strategy and has produced a new one by a peculiar form of local search. The trader has no chance of an *ex ante* evaluation of the new strategy (as is the case in the Nelson and Winter models). However, after a month of practising the new strategy, he makes an *ex post* evaluation. If we assume that the trader has remembered the 'competitiveness index' of the strategy he had before the innovation, he has a chance of returning to the old strategy if the innovation has led to lower 'competitiveness'.

5.4.4. The multi-patterned evolutionary dynamics

The specification of the iterated Trader's Dilemma has been made in a way which makes it well-suited for computer simulations. Nevertheless, since we have not discussed the probabilities of abandoning strategies, of imitating and innovating, we are not yet ready for specific simulation runs. To do so would, perhaps, be to overdo our fable of the Trader's Game. However, the patterns emerging from such simulations are an important part of the present exercise. One solution is to present simulations where the parameters have been set to correspond to the conditions of biological evolution. These studies may be seen as an extension of Axelrod's experiments which included the above mentioned (section 5.2) computer 'tournaments' between computer-implemented strategies delivered by experts (Axelrod, 1980a; 1980b) as well as by the computer itself by means of the genetic algorithm (Axelrod, 1987). The result of these grand computer 'tournaments' is that David beats Goliath or, to be more precise, that the extremely simple and relatively friendly TIT-FOR-TAT program (the $\overline{PD}$-strategy) appears to be a winner in both types of simulations. However, Axelrod (1987, 37 ff.) concludes with formulations which give the impression that things are much more complicated, and this impression is confirmed by a large literature on the limits to TIT-FOR-TAT as an evolutionary stable strategy (Lomborg, 1993). A similar conclusion can be drawn from analytical game theory.

In the present context, we shall not develop the issues relating to the TIT-FOR-TAT strategy but instead study the patterns created by computer simulations. Here we refer to the work of Lindgren (1990; 1991)[44] which demonstrates the possibility of generating surprising patterns in the evolution of the frequencies of strategies in the iterated Prisoner's Dilemma with genetic algorithms. In Lindgren's set-up, we find no simple progress towards an evolutionary stable strategy. Instead we sometimes see a pattern of 'punctuated equilibria' (see sections 2.3 and 6.2) with shifts between periods of relative stability and periods of rapid evolution, mass extinction, as well as the emerging dominance of more complex types of strategies. Such a pattern is highly relevant to a discussion of how to reform the conditions in our fable of the traders.

Lindgren's computer simulations are made on a somewhat different basis than the present version of the iterated Trader's Dilemma. First of all, Lindgren studies strategies which take into account the behaviour of both parties (A and B) in the interaction. In the case of simple 2-bit strategies, this does not make any difference, since A decides on the basis of the last action of B. But more complex strategies take into account the behaviour of both parties (see equation 5.1): with a 4-bit strategy, A takes into account the last actions of B and A; with an 8-bit strategy, A takes into account the last action of B, the last action of A, and the next-last action of B; etc. Second, Lindgren does not develop the different kinds of imitation and innovation which have been developed in sections 5.3 and

5.4. Instead he concentrates on point mutation within a strategy as well as the doubling and halving of strategies. Third, Lindgren (1991, 299) sticks to the biologically relevant low degree of strategy change: the probability that a given bit in a person's strategy is changed during a period is 2×10^{-5}, while the probabilities that his strategy is doubled or halved are both 10^{-5}. Even if a period corresponds to a trading month in the above exposition of the Trader's Dilemma game (and even if the life-time of a trader may correspond to 600 biological generations = 50 years of 12 months), this is clearly a low frequency of strategy change. But since we are dealing with a mythological and conservative Kingdom of Mercantia, the reader may ignore this problem. Fourth, Lindgren (1991, 297) applies Axelrod's payoff matrix of the Prisoner's Dilemma (table 5.2) rather than our modified version (table 5.3) which ensures that trade will not stop because of beliefs that it may give zero utility. This modification is not fundamental, even if it to some degree shifts the relative strength of the different strategies. Fifth, Lindgren emphasises noise, i.e., that the actual behaviour in a single interaction is (with low probability) the opposite of the behaviour prescribed by the applied strategy (P is changed to D, and *vice versa*).[45]

Even with these differences in contents and emphasis, it should not be too difficult to appreciate Lindgren's results. They summarise the behaviour of a population of 1,000 individuals with different strategies with respect to the iterated Prisoner's Dilemma. In each round ('generation') of the game, each of the individuals plays a large number of times against all the other individuals. The payoffs are calculated for each player, and then the relative performance of subpopulations holding each of the strategies are determined. According to the 'fitness' of each strategy, the (increased or decreased) number of applications of each strategy during the next round is found. The number of applications of a certain strategy during the next round is thus dependent upon the frequency of all strategies during the present round. Then the same calculations are repeated for a large number of rounds. The results of such huge calculations are summarised in figures 5.2 and 5.3.

Figure 5.2 (next page) covers what may be considered as a life-time of experience of a group of traders (50 years of 12 months). They stick to the four 2-bit strategies, and initially each strategy is applied by the same number of traders. The first strategy to succeed is $\overline{DD}$ (ALL-D) which exploits $\overline{PP}$ (ALL-P) and $\overline{DP}$ (ANTI-TIT-FOR-TAT). The reasons for both success and failures can be found in table 5.5. As a result $\overline{PP}$ and $\overline{DP}$ become extinct. However, the increased density of $\overline{DD}$ implies a weakening of the 'fitness' of this strategy, since it functions badly when applied against itself. Instead $\overline{PD}$ (TIT-FOR-TAT), which is initially slightly weakened, becomes successful since if cooperates well with itself, while $\overline{DD}$ becomes rarer and rarer until it disappears.

178 *Evolutionary Economics*

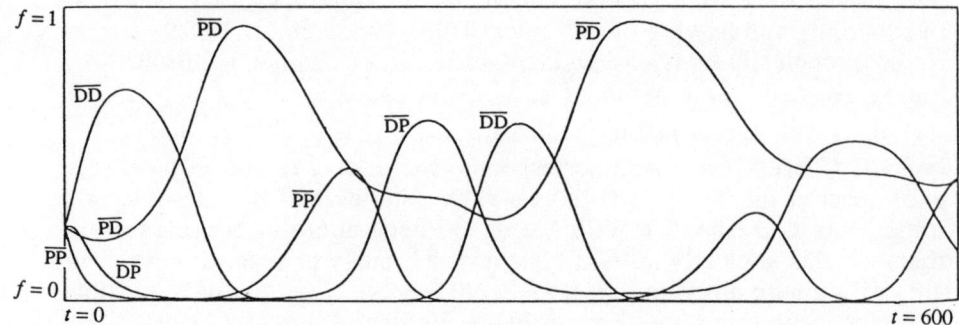

Figure 5.2. The evolution during 600 periods of the frequencies of the four 2-bit strategies (Lindgren, 1990; 1991, 302).[46]

In the new $\overline{PD}$-dominated environment there is no selection against $\overline{PP}$ which regains a certain position. However, $\overline{PP}$ can be successfully exploited by $\overline{DP}$, which in turn is exploited by $\overline{DD}$. During the last part of the 600 rounds of trading, the story begins to repeat itself with a successive introduction of $\overline{PP}$, $\overline{DP}$ and $\overline{DD}$. But all in all, $\overline{PD}$ appears to be a kind of overall winner which is able to uphold a large (but oscillating) frequency throughout the simulation.

In the longer term (e.g., '30,000 months' or '2,500 years'), this story is changed by the emergence of strategies which take into account a longer history of exchange between two parties. As earlier mentioned these 4-bit, 8-bit and 16-bit strategies (etc.) take into account the behaviour of both parties (A and B) in their process interaction. In the case of a 4-bit strategy, A takes into account the last actions of B and A. This means that we have to reinterpret the names of strategies. Thus, the $\overline{PDDD}$ strategy now means that the A will deliver a perfect good provided that both B and A have delivered perfect goods in their last interaction; in all other cases, A will deliver a defective good. Similarly, the 8-bit $\overline{PDDDDDDP}$ strategy means that A takes into account the last action of B, the last action of A, and the next-last action of B. A will deliver a defective good unless all of these goods have either been perfect or defective.

The overall pattern of the simulation of a large number rounds depicted in figure 5.3 may be described as 'punctuated equilibria' where periods of stability in the frequencies of strategies are interrupted by periods of unstable dynamics. During the 30,000 periods we see the emergence of four epochs of stasis. In the first of these epochs there is a coexistence of $\overline{PD}$, $\overline{DP}$ and a low frequency of 4-bit strategies. For thousands of periods only small changes in frequencies occur, and during this epoch the experts of the Kingdom of Mercantia would probably think of the system as being in an equilibrated state.

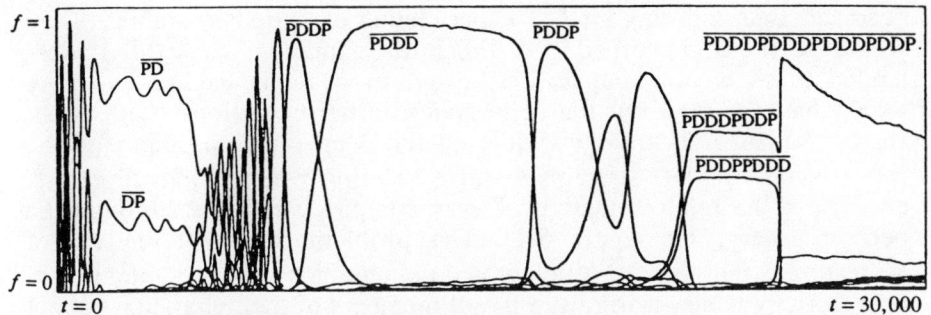

Figure 5.3. The evolution during 30,000 periods of the frequencies of strategies using an increasing amount of memory (Lindgren; 1990; 1991, 303).

Nevertheless, a new instability emerges. Luckily, this is followed by a new period of a high degree of stability with a very high frequency of a special version of the TIT-FOR-TAT strategy, namely $\overline{\text{PDDD}}$ which react against B's as well as A's defective goods. One might think that this is the end state and that $\overline{\text{PDDD}}$ is an evolutionary stable strategy. However, after another turbulent period, the 4-bit strategies become almost extinct and the epoch of the 8-bit strategies begins. Here we see a coexistence of two strategies which do not function well in isolation. The reason is that they cannot cope with an accidental defection when trading with each other; but when trading against each other, the two strategies are able to overcome an error (Lindgren, 1991, 305). Thus, we see that in iterated games involving both strategies, an error will only lead to a short period of defection after which cooperation is reestablished. This may be taken as a primitive form of mutualism and the simulation results show many other forms of coevolution. But in the last epoch depicted in figure 5.3, the dominant 16-bit strategy is able to collaborate with itself, even in the case of accidental errors.

Through other large-scale combutations, Lindgren (1991) has demonstrated that this strategy is something like a relatively broadly applicable evolutionary stable strategy. Lomborg's (1993) study of iterated Prisoner's Dilemma games, with an alternative (misconception-oriented) type of noise, generates a more pluralistic type of the metastable cominance of cooperation. Thus we come close to an ultimate stasis which was searches for by Smith (1982) and Axelrod (1984). The complexity of this/these solutions is, however, dependent upon the fact that information about previous interactions is a free good in the games (e.g., supplied by our mythological notary) and that the storing of longer and longer strategies is also costless. If these assumptions are removed, then a simplistic strategy like TIT-FOR-TAT may once more become attractive, although it is not a formally proven evolutionary stable strategy.

The experts advising the Queen of Mercantia would probably not have been satisfied with this kind of experimental evolutionary studies. They would be more concerned with the initial situation where all traders tended to show non-cooperative behaviour. On this background they would have pointed out that Lindgren's initial conditions (with equal shares for several strategies) missed the Mercantian problem of the pervasiveness of $\overline{DD}$-strategists and a solution which removed errors, even from the implementation of very complex strategies (through the perfect notary, see figure 5.1). The problem for the Kingdom of Mercantia is that the $\overline{DD}$-strategy is an evolutionary stable strategy, that is, a strategy which if followed by all members of the population cannot be invaded by a single innovator (mutant). This conclusion may be arrived at through table 5.4 and simple reflections about the outcome of all possible strategies. A *single* application of any other strategy will always be beaten.

This result appears to be quite fatal in relation to the fable about the Queen's reform of the trading system. The only chance for creating a change is that several traders at once turn to more cooperative strategies. The reformed Trader's Dilemma is set-up in a way which makes this situation probable. However, other ways of reformulating the game would make change even simpler. The task is to favour mutual transactions between cooperatively oriented strategies to a degree which will outweigh the losses from transactions with $\overline{DD}$-strategists. One possibility is to introduce errors in the implementation of strategies or the interpretation of the quality of thre goods delivered. Another possibility is to let the innovators and the first of their imitators make more transactions with each other than with the rest of the trading population; in this way the advantages of the cooperative strategies can be exploited relatively quickly. Nevertheless, this attempt to bring 'structure' into the population P actually subdivides it into two or more populations, $P_1, P_2, ...$ which are more or less segregated. One might even consider the situation where the population splits into two, totally seggregated populations. These are quite important problems in the discussion of evolutionary processes but they clearly transcends the limits of the present book as well as the limits of the story of the Mercantian traders.[47]

Notes

1   In this respect a recently reprinted paper (Simon, 1956/92) is typical: Simon presents a model of a simplified organism with the single goal of obtaining food in a complex environment where resources are unevenly distributed. There are different proposals for the extension of the model, like a more structured environment and multiple goals of the organism. However, the possibility that there might be other organisms in the environment is not mentioned with a single word. Thus, we have yet another version of Robinson Crusoe on a desert island.

2   Who also, together with Simon and Shaw, developed many other ideas and tools of Artificial Intelligence.
3   See any textbook in Expert Systems or Artificial Intelligence, e.g., the short version in Rich (1983, 31 ff., 48 ff.). Shorter accounts of production systems are found in Simon (1981, 20-123) and Haugeland (1985/89, 157-164).
4   The symbol ___ denotes a conditional rule rather than a list of situational information. Later in the chapter 'P' and 'D' is used instead of '1' and '0'. Apart from notational convenience, the change has been made to emphasise that we are later describing complex strategies rather than simple conditional parts of rules.
5   In Germany, Rechenberg introduced so-called Evolution Strategies for solving complex optimisation problems. Hoffmeister and Bäck (1991) have demonstrated the important similarities between this approach and Holland's genetic algorithms.
6   Arthur has earlier random walk models and urn schemes for demonstrating the possibility of a lock-in of the system into non-optimising solutions; see a summary in Arthur (1988).
7   The agent's process of learning and confidence-building is described by one or two parameters (Arthur, 1993, 5 f.). In earlier versions of the model (1991, 354) only a two-parameter case was described. In my exposition the reciprocal of Arthur's single-parameter is used, since this revised parameter is proportional to the rate of confidence-building (called learning).
8   An initialisation procedure is necessary. At the beginning of the first period all brands should be given an arbitrary but positive strength, the vector of strengths should add up to β, and a brand should be selected on this foundation.
9   The functioning of the algorithm can be demonstrated by a simple example where an agent in each period has a choice between two brands, A and B. Assume that he initially has an equal probability of choosing the two brands, but that he by chance chooses B two times. Assume, furthermore, that the value of B is judged to be 4 units in each round. If $\alpha = 4$, then both A and B must initially have had the strength 2 (2 + 2 = 4). After the first round the figure of B is increased to 6 (= 2 + 4), and renormalisation leads to strength 3 (and 1 for A). After another try, the strength of B is 3,5 (and 0,5 for A). Thus, the probability of choosing B in the next round is 0,88.

If instead $\alpha = 16$, then both A and B must initially have had strength 8. After the first round the figure of B is increased to 12, and renormalisation leads to strength 9,6 (and 6,4 for A). After another try, the strength of B is 10,9 (and 5,1 for A). Thus, the probability of choosing B in the next round is 0,68.
10 Some subsequent computer simulations study the changes in behaviour and the closeness of the long-term behaviour to the optimal one (Arthur, 1993).
11 Although the results of the work of, e.g., Arthur is important, it is obvious that Artificial Economic Evolution should avoid the kind of overselling of premature results which is so well-known from the history of Artificial Intelligence—not least in Simon's promotion of it (see Dreyfus, 1972/79, 81-85, 93-96).
12 The chapter concentrates on the creation and analysis of iterated games which are oriented towards computer simulation. This means that it does not cover the study of the evolution of repeated games by analytical game theory (see, e.g., Friedman, 1992). Even within the experimental approach, there are many contributions which cannot be taken into account. In this way the chapter follows the same strategy as chapter 4 with its emphasis on Nelson-and-Winter models. The present chapter covers Axelrod-type models, while several parallel developments are more-or-less ignored. For instance, Hirschleifer (1982/93) has independently developed many of the same thoughts as Axelrod, see Hirschleifer, 1987, 211. He also applies other games than the Prisoner's Dilemma.
13 However, one may , e.g., rethink the model with entry to the industry (section 4.4.2) in terms of games. A firm with a large market share may, as suggested in section 4.5, use overcapacity and high search costs to create barriers to entry.

However, the introduction of these ideas presupposes major changes *vis-à-vis* the typical Nelson-and-Winter models.

14 This table is a slightly modified version of a table in Winter (1991, 187).
15 The emphasis on the two-person Prisoner's Dilemma is necessary to simplify the following study. However, the *n*-person Prisoner's Dilemma is quite important in many social situations with 'free rider' problems. An evolutionary analysis of this game would be of great interest.
16 See Axelrod, 1980a; 1980b; 1984; survey in Hofstadter, 1985b.
17 In the second tournament a variable length of the iterated games was used in order to avoid 'end game' tactics.
18 See Axelrod, 1984, 8. Payoffs to player A are listed first. Thus, the value for A of a game where both cooperate is 3 while the value for A of a game where A is cooperative and B defects is 0. More formally: $V_A(C|C) = 3$ while $V_A(C|D) = 0$. Axelrod's (1984, 8) choice of a payoff matrix is just one of an infinity of possibilities. The general structure of the problem is well-known. For player A the conditions are:

$$V_A(D|C) > V_A(C|C) > V_A(D|D) > V_A(C|D) \text{ and } V_A(C|C) > \frac{V_A(C|D) + V_A(D|C)}{2}.$$

Similar conditions should hold for player B but the matrix does not have to be symmetric.

19 The simulation model (Axelrod, 1987, 36-38) is built on specific assumptions which are of little interest in the present context. However, these assumptions may give an impression of the procedures used. Axelrod operates with a population of twenty individuals. Each of them has a fixed strategy which takes into account the outcomes of precisely the three previous moves (64 different histories). In order to get the game started, Axelrod has to introduce some hypothetical moves (since the strategies need information about the last three moves). In all, this leads to an overwhelming number of possible strategies (2^{70}). Each individual plays with 8 other individuals in each generation. Each of these plays consists of 151 moves. The composition of the next generation reflects the relative success of the programs. In each generation 1/2 mutation and 1 'crossover' of the 'genotype' is made in a stochastic way. One computer simulation consists of 50 generations. Eighty simulations are made, half of them with 'sexual', half with 'asexual' reproduction (depending on whether the 'genotypes' between two parents are mixed or whether each individual has only one parent). One of the weaknesses of this simulation is that it only allows strategies which are based on the three last moves of the other player. A more flexible approach is used in the present chapter.

20 It should, however, be noted that there are other possibilities of applying the genetic algorithm and evolutionary games. For example, Marengo (1992) has tried to demonstrate that an extended version of the genetic algorithm is well-suited for an experimental study of Aoki's (1986) hypotheses about the contrast between the organisational structures typical of Japanese and American companies, especially with respect to intershop coordination problems. Marks (1992) summarises an attempt to go from the simple Prisoner's Dilemma to a Generalised Prisoner's Dilemma for studying the pricing behaviour of oligopolists using a kind of TIT-FOR-TAT strategy; furthermore, he tries to show how the genetic algorithm can be used for generating strategies for this game and thus give some indications of the emergence of the (relative) evolutionary stability of the successful strategies. Friedman (1991) gives a survey of the larger area of evolutionary games in economics with an emphasis on the formal aspects of the studies. Sugden (1989/93) exemplifies the growing interest in the economics of institutions which emphasises the emergence of institutions as a kind of spontaneous order; for the study of such processes the Prisoner's Dilemma is supplemented by other types of game (e.g., 'hawk' and 'dove' strategies in the Chicken game).

21 The supply side of the story will not be included in this discussion. However, the reader may envisage a case where the production costs of a defective and a perfect commodity are (nearly) the same.
22 In comparison to table 5.2, the payoffs have been changed in order to make them all positive. The positive outcome gives the traders an incentive for proceeding with the game without taking into account their evaluation about the possibilities of ending up in the worst possible outcome of the game.
23 There are two possibilities to avoid a potentially expanding 'end game' of cheating behaviour. First, it may only be the average number of transactions between two traders which is made public. Second, the probability, w, that a given transaction between two parties will be followed by a new transaction may be less than one.
24 To avoid tactical behaviour in the traders' information to the notary, they are supposed to be kept in ignorance about the existence of the tax index.
25 To simplify, we operate with a fixed array instead of a variable-length list. In order to avoid the end-game situation the array may, e.g., include the double of the approximate number of transactions during a period (month).
26 In his first exchange in a new month, A has no information about the behaviour of B. But later in the month he may examine the records. To generalise we assume that A is always looking at the records.
27 An ineffective variant of the strategy can be created by delivering a defective commodity in the first move.
28 The moves of A in rounds 1 and 2 are not specified. In these rounds he delivers perfect commodities. However, alternative 'opening games' will create variants of $\overline{PDDD}$.
29 See definition 5.2, the formats of definitions 5.3 and 5.4 as well as the text.
30 Payoffs to trader A are listed first. Each cell of the table specifies $\left(V_{100}^A\left(\overline{X}|\overline{Y}\right), V_{100}^B\left(\overline{Y}|\overline{X}\right)\right)$, without noise, discounting, etc.
31 In the case where the mistake takes place in the last transaction, there is no consequence and thus the full potential benefits will be reaped by the $\overline{PP}$-strategist. Lomborg (1993) has dealt with more general problems; he has especially made simulation experiments with different levels of noise in the iterated Prisoner's Dilemma.
32 To simplify the following interpretations we assume that we are dealing with a set of supergames with no discounting and no noise.
33 However, a notary brought up in the book-keeping tradition would not dare to follow this aggregative procedure. She would have to handle information on each trader individually in her records.
34 Of course, $\sum_{j=1}^{\overline{m}_t} N_j = p$.
35 This and the following definitions are formulated as assignment statements in the same style as used in chapter 6, see also appendix 1. This notation has been chosen to emphasise the computational aspects of the study of the Trader's Dilemma game.
36 For convenience we operate with a constant number of traders, p, within the population of traders, P. We could, however, generalise to a situation where the population consists of a varying number of traders.
37 Otherwise an evolutionary process would not take place, see below.
38 The notion 'choice' is here used in a broad sense which potentially (but not at present) may include 'choice' of strategy by random errors or 'mutations'.
39 And still fewer possibilities of allowing a to vary over time, etc.
40 In a more general context this problem has been pointed out by Winter, 1987, 614.
41 This operation is not the same as Holland's version. He tries to simulate sexual reproduction where a 'crossover' sometimes takes place between the genetic strings from the two parents. In Holland's version a split is made at the same randomly determined point in the parents' genetic strings, after which the parts cut off are swapped.

42 This possibility is not included in Holland's set of idealised genetic operators. It is, however, of much interest to the exploration of the iterated Trader's Dilemma since it permits in one step changing the strategy from considering the last n moves to considering the last $n + 1$ moves. For reasons of implementation we need to put an upper limit to the length of strategy strings created by repeated doubling; for example, we may define a 128-bit string as the upper limit (which means that 7 moves are taken into account). An alternative approach is to make random mutations all along a 128-bit string which initially may, e.g., consist of $\overline{PD\#\#\#\#...\#}$, and which is changed to $\overline{PD\#\#\underline{D}\#...\#}$ after the first mutation. However, while this is perfectly rational when the task is to use the genetic algorithm to create approximations to globally optimal strategies, the procedure is awkward in terms of simulating evolutionary processes.

43 Some implementation issues and available programming systems are presented in Stein, 1989.

44 The simulation uses Axelrod's payoff scheme, the genetic algorithm with a small propensity to innovate, and random errors in the application of strategies. See also Emmeche (1991, 104-108, 190 f.) and Lomborg (1993).

45 This is only one of the two major types of noise in the environment (Axelrod and Dion, 1988). Apart from this misimplementation by an agent of a strategy, we have also the misperception of the action of an agent by another agent. Lomborg (1993, 8) argues that the latter form of noise is most relevant, and thus that Lindgren's results are based on a problematic assumption. However, even in computer-simulated iterated Prisoner's Dilemma games with the misperception type of noice, he reaches the conclusion 'that to a remarkable degree, cooperation is possible, fairly stable, and will reach a counterintuitively high score that is much more efficient that TIT FOR TAT, even under high levels of noise.' (p. 20)

46 Lindgren is using another notation for strategies, in the style of Smith (1982). In Lindgren's notation, the reactions against defective actions are listed before the reactions against perfect actions. Furthermore, transaction $i - 1$ is listed before transaction i. In the present chapter, both procedures are reversed (see table 5.4). In figures 5.2 and 5.3, which present the results of Lindgren, the notation of the present chapter has been applied.

47 Axelrod (1984, 65-68, 98-100, 213-215) has started considering the possibilities of creating a cluster which trades more internally than with the hostile outside world of trade.

6. Research horizons

6.1. Towards an evolutionary-economic synthesis

The history of science demonstrates that research work does not always progress directly from simple to more complex issues. In the initial developments within many research areas, the progress is rather characterised by the reverse movement from complex to simple tasks. Some of the grand ideas from the early days of scientific work have to be pushed aside before researchers can start to construct a solid foundation for a long-term effort in their chosen area of investigation. However, the visions of the pioneers may still serve as a reminder of the remaining issues, as a 'demand specification' of what the discipline should ultimately be able to cover. An important example of such a development is found in biology (see Provine, 1971; Mayr, 1982; Mayr and Provine, 1980). Here Darwin's (1859/1964) *On the Origin of Species by Means of Natural Selection* integrated results from an enormous area including such diverse subjects as, e.g., systematics, biogeography, and palaeontology. However, some biologists had to retreat from the grand scheme in order to develop an understanding of the precise mechanisms of inheritance as well as to model formally the evolutionary process. In theoretical population genetics the foundation was created which helped the subsequent general development of evolutionary biology. In retrospect, one may say that a convenient division of labour emerged between certain theorists who emphasised microevolutionary phenomena within a given population and more practically oriented naturalists who dealt with the macroevolutionary phenomena of interaction and segregation between different populations. Nevertheless, while the novel work was performed in the first decades of this century, it tended to isolate itself from naturalists interested in the same macroevolutionary phenomena as had been Darwin's starting point. Furthermore, the theoretical and laboratory-oriented work also created many controversies, e.g., about gradualism and punctualism, which in retrospect appear to be mistaken. Fortunately, some biologists were able to bridge the gap between the clear-cut analysis of microevolution and the macroevolutionary phenomena (like speciation) which interested practically oriented naturalists. In this way, such biologists (like Dobzhansky, Huxley, and Mayr) created the evolutionary synthesis on which much of modern biology built.

When comparing old and new evolutionary economics, we clearly see the movement from very complex to less complex phenomena, and from macroevolution to microevolution, i.e., from the level of several interacting 'populations' of rules to the level of the individual 'population' of rules. In relation to Schumpeter's core case of 'railroadization', we may say that the focus has been moved from the overall picture to the analysis of very limited parts of the picture like the relative performance of different kinds of tracks (of cast iron or steel). We also see a shift away from work based on statistics and qualitative historical data toward computer-based experiments with Artificial Economic Evolution. Seen from the viewpoint of Schumpeter's 'demand specification' for evolutionary economics (see ch. 2), this development is both bad and good. It is bad because it means a retreat from the macroevolutionary phenomena and from the punctualist vision which Schumpeter represented. It is good because analytical tools are now provided which Schumpeter was so desperately missing. It is also good because many researchers have followed Schumpeter's 'final thesis' on the need for a large collection of historical case studies in order to provoke and guide theorists. However, it must be admitted that large parts of the theoretically oriented new evolutionary economics from Alchian to Arthur have only had a loose empirical orientation and a weak relationship to the old evolutionary economics. It is rather the population thinking and the algorithmic description of evolutionary mechanisms which have been its driving forces.

These mixed conclusions may, perhaps, indicate that it is not very fruitful to make a direct confrontation between Schumpeter's 'demand specification' and the results which have until now been delivered by new evolutionary economics. Such an immediate evaluation of the new results is bound to create unnecessary controversies and misunderstandings. In this way we appear to repeat some of the mistakes which evolutionary biology made in the first decades of the century. Instead, we should evaluate the contributions in relation to a larger evolutionary-economic research programme in which the most difficult tasks are connected to the explanation of macroevolutionary phenomena—like the diversity and adaptedness of the institutions and structures of economic life—and to the interaction between the evolutionary process and the movement of macroeconomic aggregates. There is no reason to demand that evolutionary-economic research confronts such explanatory tasks directly. Actually, it would be a bad research strategy to do so. The *Magna Carta* of evolutionary economics would rather consist in the demonstration that a primitive process of economic evolution can be synthesised from economically relevant mechanisms concerning the preservation and transmission of rules, the creation of new rule variants, and the selection between them. In this respect, real progress has taken place from Alchian's and Winter's initial suggestions via the Nelson-and-Winter models of the process of 'Schumpeterian competition' to recent attempts in Artificial Economic Evolution. This development has led to an increased understanding of the surprisingly difficult concept of an evolutionary process as well as to the exploration of important characteristics of evolutionary processes like, e.g., the importance of the timing of events as well as the stickiness in responses of economic agents.

From the viewpoint of the development of evolutionary economics as a long-term process, the contributions of new evolutionary economics are quite crucial. Nevertheless, the question is now whether we can expect from it a steady progress towards the macroevolutionary phenomena which were its initial motivation. The answer is that there is no clear tendency in this direction. On the contrary, we may expect that the difficulties involved in attempts to conceptualise and model even simple evolutionary processes (see chs. 3-5) will discourage researchers vis-à-vis the more ambitious tasks. Furthermore, the shift to a macroevolutionary level will imply the introduction of new mechanisms whose importance cannot be seen from a microscopic point of view. Instead of expecting that the specialists in microevolutionary analysis will automatically shift to deal with the broader tasks, there is a need to educate researchers in a way which makes them more able to bridge the gap. New competencies are needed in order to create a modern synthesis in evolutionary economics of the same kind as the biologists created in the 1930s and 1940s. At that time the term 'evolutionary synthesis' was introduced to designate the combination of formal analysis of the evolutionary microlevel and theories dealing with macroevolutionary phenomena. Here a whole new set of concepts like 'isolating mechanisms', 'sympatric and allopatric speciation', 'founder principle', 'gene flow', 'isolate', 'stabilising selection', 'taxon', etc. emerged.

The creation of a new evolutionary terminology greatly contributed to the eventual synthesis. At least some of the misunderstandings resulted from the lack of an appropriate and precise terminology for certain evolutionary phenomena. Nearly all the architects of the new synthesis contributed terminological innovations. (Mayr and Provine, 1980, 29)

The present book has been designed as a stepping-stone on the difficult way towards a similar modern synthesis within evolutionary economics. In this perspective it is the contrast rather than the unity of the book which should be emphasised. On the one hand, we have Schumpeter who through a confrontation with Walras and other marginalist economists developed a magnificent but crude picture of the macroevolutionary phenomena and their relationships with basic economic categories like profit, interest, and business cycles. On the other hand, we have the new evolutionary economists who have mainly clarified the notion of an evolutionary process (chapters 4-5) but who have also taken the first steps in describing and analysing macroevolutionary phenomena (chapter 2-3). If these diverse elements stay together, they appear to be more suited to start the development of a modern evolutionary-economic synthesis than if formal evolutionary analysis departs from the Schumpeterian heritage.

That is the basic message of the present book and some work to substantiate it has been presented (especially in chapters 2-3).

On the way towards a modern synthesis it is important to accept the multiplicity of evolutionary economics. Especially, the four characteristics which were presented in the introduction to the book will, in my opinion, be important for a viable new evolutionary economics, namely: its application of population thinking, its empirical orientation, its use of an algorithmic approach to clarify evolutionary mechanisms, and its 'dialogue' with the classic works on economic evolution. In practice things have shown up to be more complicated: empirically oriented researchers often doubt the relevance of population thinking and algorithmic approaches while theoretically minded economists would at least suggest an arms-length approach to historical studies and to classic authors. The book has tried to take an intermediary stand. It has primarily been designed in relation to new evolutionary economics but it has tried to define important relationships with Schumpeter. It has been primarily theoretical but it has emphasised the empirical relevance of the theoretical frameworks as well as their contribution to concrete computer models. The book has, furthermore, suggested that the case of 'railroadization' is not only an important illustration of Schumpeter's ideas but also a source of inspiration for the further development of evolutionary economics.

This mediating strategy has found a good deal of support in the literature covered by the book. Schumpeter is undoubtedly a major discussion partner for much of the work within new evolutionary economics. The empirical orientation is clear, even in the most abstract contributions like the evolution of strategies toward the iterated Prisoner's Dilemma. In the end, economic evolution depends on such historical 'details' as the propensity to stick to rules, the speeds of innovation and imitation, and the structure of the relevant search spaces. Despite this support to the argument, there may still be some doubts about the central theses of the book. Couldn't we just as well say that the post-war development of new evolutionary economics has falsified my own thesis of the need for a 'dialogue' between new and old evolutionary economics as well as Schumpeter's 'final thesis' on the need for a large collection of historical case studies as a means to promote the understanding of economic cycles and evolution? At least it seems as if the theoretically oriented new evolutionary economics has developed its contributions through abstract modes of population thinking and through formal descriptions of evolutionary mechanisms rather than through any clear-cut historical orientation. Furthermore, most of these contributions have been developed without any explicit dialogue with old-fashioned evolutionary approaches like that of Schumpeter. Finally, the new clarity about the character of evolutionary mechanisms seems to open up a huge research programme; in other words, our complex evolutionary mechanism can account for a large number of different processes. Why should we under such circumstances restrict the application of these mechanisms to any historically oriented definition of the character of the evolutionary process or to any defunct theorist's ideas?

The answer is: because 'the subject matter of economics is essentially a unique process in historic time.' (Schumpeter, 1954, 12). This subject matter of the discipline is, furthermore, defined by the history of economic thought. This means that economic history and the history of economic thought define the subject matter of the discourse in a way which cannot be ignored without a risk of confusion about the direction and the meaning of the economic argument. Instead, we should recognise that the history of evolutionary phenomena as well as the history of economic thought help us to answer the apparently simple question: 'What evolves?' Contrary to its apparent simplicity, this question is really the most fundamental and most difficult question of evolutionary analysis, which even today has not been answered in a fully satisfactory way. The question is so important that we repeat it in an unforgettable way which is recorded by Boulding, a contributor to post-war non-formalised evolutionary economics, who became fascinated by Schumpeter and economic evolution while he was a pre-war student at Oxford University:

My Oxford philosophy tutor, who had the curious habit of crawling under the table while giving his tutorials, commented in a high British voice coming from underneath the table on a paper I had given on evolution, 'It is all very well to talk about evolution, Mr. Boulding, but what evolves, what evolves, what evolves?' (Boulding, 1978, 33)

Forty years after this conspicuous form of pedagogics, Boulding had a 'glimmering' of an answer: 'What evolves is something very much like knowledge.' (*ibid.*) While this answer is undoubtedly correct, it is also radically incomplete in relation to the development of an analysis of economic evolution. Especially, we would like to find an evolving substance which has a much less amorphous character than the commonsense kinds of 'knowledge'. To be able to give rise to an evolutionary process, the 'thing' we are studying should have an aspect of preservability, mutability and selectability. Evolutionary biology started with an apparently much more specific answer ('species'), but evolutionary economics has normally avoided most of the discussion raised by Boulding and his tutor. Instead, the normal strategy has been to start with a certain type of explanation and then to search for something which needs to be explained: instead of defining an analogue to the *explanandum* (to the species), the dominant tendency in evolutionary economics has been to define an analogue to the *explanans* (in Darwin's case: natural selection). By postponing the crucial answer to the question 'What evolves?', we tend to leave open the subject matter of evolutionary economics. We only suggest a general mechanism for the explanation of practically everything just like the neoclassical economists do in relation to non-evolutionary matters. We tend to develop a framework 'On the Origin of X by Means of [a Well-Defined Mechanism]' rather than 'On the Origin of [a Well-Defined Phenomenon] by Means of Y'. The empirical orientation and the emergence of a modern synthesis will, hopefully, help us to avoid this danger of ending up in a Hegel-like evolutionism which 'explains' everything.

6.2. The diversity of evolutionary explanations

This book's exposition and development of contributions to post-Schumpeterian evolutionary economics have demonstrated two central and interconnected problems: first, evolutionary explanation is quite complex and raises a series of methodological and theoretical difficulties; second, several modes of evolutionary explanation have emerged which provide different solutions to the difficulties—and these solutions are partly complementary and partly competing. From an evolutionary perspective the second point may in a double sense be seen as an answer to the first: on the one hand, we may have a variety of trials-and-errors which in the end may come up with an answer to the difficulties of evolutionary-economic analysis; on the other hand, we may face a relatively permanent diversity of approaches which reflects different aspects of economic evolution. In biology the tendency have been that many of the original approaches have disappeared while others have been integrated in the neo-Darwinian synthesis. The situation is more open at the present stage of the development of evolutionary economics. There are obvious possibilities of creating a modern evolutionary-economic synthesis. But this does not necessarily lead to a removal of the different approaches which may still survive because they provide explanations to different types of evolutionary phenomena. For obvious reasons evolutionary researchers tend to specialise in a selected area of expertise. But some knowledge of the overall situation is important for would-be contributors to a modern evolutionary-economic synthesis. Here, the individual biological approaches often serve as points of reference for describing the evolutionary-economic approaches.

6.2.1. The past multitude of biological explanations

It is quite interesting to review the relatively mature situation within evolutionary biology and especially the historical development which has led to modern forms of evolutionary-biological explanation. The invocation of this material in a treatment of evolutionary economics is based on several structural analogies or, perhaps, isomorphisms between the theories of biological and economic evolution. It is not being suggested that evolutionary economics is based in biological analogies or that I propose a general application of such analogies (criticised by Schumpeter, 1912/34, 57 f.; 1954, 789; Penrose, 1952). Analogies may be important as a 'fertile source of new ideas and explanations' but 'it is important not to get carried away by evolutionary analogies and to mistake the analogy for the reality' (Freeman, 1992, 123). In the end the evolutionary-economic results must be justified on their own grounds as was emphasised in Alchian's answer to Penrose's critique of his use of the analogy:

The theory I presented stands independently of the biological analogy. ... In my original article every reference to the biological analogy was merely expository, designed to clarify the ideas of the theory. ... Readers of an earlier draft, containing no references to the biological similarity, urged that the analogy be included as helpful to the understanding of the basic approach. (Alchian, 1953, 601)

Even in the present context, biology may help to increase the clarity and distinctness of different modes of explanation, especially by demonstrating the historical existence of different types of evolutionary explanation.[1] To make the exposition more concrete and to emphasise its informal character we take up a classic question: Why do giraffes have long necks?

1. Functional explanation points out that this attribute (long necks) is quite helpful in certain environments like the savannah. This explanation is, however, not sufficient since the next question is: which mechanism creates functionally adapted animals? Provided we have a widely acknowledged mechanism of evolutionary adaptation (natural selection), we may, however, consider a functional explanation as a legitimate short-hand form of a full explanation (Elster, 1983, ch. 2).

2. Teleological explanation (Mayr, 1982, 47-49, 372 f., 528-530) includes the specification of a mechanism formulated in terms of goals and foreknowledge: giraffes are considered to be (Divinely) designed rather than outcomes of an evolutionary process. In somewhat updated terms we may summarise pre-Darwinian teleology in the following way: God wants a harmonious world; to create a global optimum in this world, he has placed long-necked giraffes in it. However, the main point of this approach was not to explain nature but to prove the existence of God by reference to the existence of cases of adaptation in nature (like the neck of the giraffe).

3. Orthogenetic explanation (Mayr, 1982, 528-531; Bowler, 1983, ch. 7) suggests that species evolve linearly but in a non-adaptive way. The evolution of a species is not seen to be driven by any adaptation to the environment but instead by an organism-intrinsic mechanism which gives direction to the evolution of species. The background for this view was that palaeontologists had noticed very long-term evolutionary trends for which there were no apparent environmental explanations. If this type of explanation was ever applied to the giraffe problem, it must have suggested that the long neck evolved because of a mechanism based in the (genetic) constitution of giraffes. No optimal properties are suggested unless a selection mechanism forces adaptation to take place or a Divine plan is supposed to be placed in the genes.

4. Lamarckist explanation (Bowler, 1983, ch. 4; Mayr, 1982, 343-360) only reflects one aspect of Lamarck's complex explanatory model—the inheritance of characteristics acquired during the life of individual

giraffes. Focusing on the use-inheritance mechanism, we may say that giraffes strive for the abundant leaves of the trees and thus develop longer and longer necks; when the necks of giraffes stop growing, a locally optimal solution has been found. We do not know whether this solution is globally optimal since learning-by-doing is incremental and since the coevolution of giraffes and trees might have created a situation where there is little extra food to find immediately above the reach of giraffes but plenty of food a couple of metres further up.

5. Saltationist or mutationist explanation (Mayr, 1982, 542-550; Bowler, 1983, ch. 8), which was dominant around the turn of the century, is based in the belief that there are two types of evolutionary change: gradual change within existing species and jump-wise formation of new species. This distinction (of Vries and others) is founded in the belief that normal individual variability cannot transgress the borders of species. New species, like giraffes, emerge because of the spontaneous and sudden production of a discontinuous variant (with a radically longer neck than its fellow animals). In contrast to the orthogenetic account this jump is seen as basically random. Since such saltations are rare and appear without any visible preparation, we cannot assume an encompassing testing of alternatives and it, therefore, is not appropriate to talk of the optimality of the new solution.

6. Darwinian (and Darwinian-Mendelian) explanation (see, e.g., Mayr, 1982, 477-488) gives a non-intentional account for the long neck in terms of natural selection: Species show hereditary differences. At a certain stage of the evolutionary process the hereditary attribute 'long-neckedness' gave lucky proto-giraffes comparative advantages. In probability terms the individuals with longer necks created more progeny than short-necked individuals. The cumulative effect of selecting the fittest members of the species for many generations will be a locally optimal solution. Here the length of the necks of giraffes still varies but the mean length of necks is much larger than in the initial situation.

The long-standing and on-going controversies within and around evolutionary biology[2] serve to emphasise some of the difficulties in coping with evolutionary explanation which are of great interest to evolutionary economics. However, the biological debate has ended with Darwinian-Mendelian explanation as the clear winner. The functional mode of biological explanation is not a real competitor: it does not provide a self-sufficient explanation because it lacks direct reference to a causal mechanism; thus it can be considered as a shortened version of Darwinian explanation. The teleological explanation is ruled out since it is not scientific because of its reference to a mechanism which cannot be tested. The orthogenetic explanation turns out to be extremely loose and it placed the main explanation in undescribed characteristics of the genes. The remaining modes of explanation are formally acceptable but it has become clear that there is no Lamarckist mechanism (except in insignificant cases).

The final choice between saltationism and Darwinism was not easy to make at the turn of the century when Mendel's laws of heredity were rediscovered: to many researchers the mechanisms of heredity appeared sufficient to explain evolution without reference to selection and the survival of the fittest. This led to an eclipse of Darwinian explanation and gave credibility to the saltationist view (Provine, 1971; Bowler, 1983). However, the emerging use of mathematical modelling of evolution (by R.A. Fisher, Haldane, etc.) demonstrated that Darwinism and Mendelism were perfectly compatible and that small selection pressures could create the great effects by the laws of exponential growth of variants (originally suggested by Malthus). There appeared to be no need for the imprecise saltationism. In combination with concrete theories of the processes of speciation, the mathematical models developed into the modern evolutionary synthesis, or the synthetic theory of evolution (Mayr and Provine, 1980). Today, many think that there is only need for one type of evolutionary explanation. However, a partial revival of saltationism has been made in order to take account of what appear to be long-term shifts between stasis and rapid change in the evolutionary process, especially connected to the creation of new species and higher taxonomic units. The modern theory of 'punctuated equilibria' (Gould and Eldredge, 1977; Eldredge, 1989) which we met in chapter 2 is, however, much more closely connected to the basic Darwinian mechanism than to turn-of-the-century saltationism. This relationship is made clear when we reconsider the species-creation part of the modern evolutionary-biological synthesis which deals with the discontinuities of evolutionary change. The terminology created here includes 'isolating mechanisms', 'the founder principle', 'stabilising selection', etc. (see above).

8.2.2. The diversity of evolutionary-economic explanations

When we turn to economic explanations, we also find a dominant type, but this neoclassical form of explanation has traditionally not only ignored evolutionary processes but also excluded them from the analysis because of the assumption of fully informed decision-making with infinite powers of computation. Such an analysis cannot accept the mechanism-oriented explanation demanded by the realist methodology and prefers instead an instrumentalist approach with its sole emphasis on the validity of the predicted outcome of the economic process. This issue was treated in the realism-of-assumptions controversy around Friedman's (1953) methodological paper. The background for the dominance of an instrumentalist rather than a realist view in economics is that we in comparison with biology have major additional complications which relate to the fact that we operate with agents whose behaviour is to some extent reflecting complex expectations about the future as well as their learning from past experience. Here we confront the fundamental problem that subjective values and constraints of the economic agents cannot be dealt with empirically in a fully satisfactory manner.

Let us consider the different modes of explanation in relation to a simple explanatory task (its formulation relates to Matthews, 1984/93): We have an economy with at least two alternative forms of behaviour, A and B, e.g., two alternative methods of production. Let it be a fact that A is persistently used by the economic agents under the given circumstances. Why?

1. The strong version of rational-choice explanation[3] says that A is chosen because (a) it is, *ceteris paribus*, the globally best alternative with respect to the agents' preferences (profit maximisation) and (b) the agents are perfectly rational and informed about the future effects of choosing A, B and all the other possible alternatives. Such agents will immediately turn to the globally optimal alternative. Weaker versions of the rational-choice explanation may also be formulated but then the subsequent equilibrium analysis becomes much more complicated. Such agents know everything about the future and go directly for the final solution without any evolutionary process.

2. The functionalist explanation appears to some authors to be the evolutionary explanation *par excellence* because they think of Friedman's above-mentioned argument. Thus Simon remarks that:

It is this view of rationality—from the standpoint of its results—that has been most prominent in evolutionary theories. Evolutionary theories explain the way things are by showing that this is the way they have to be in order for the organism to survive. *How* the organism achieves its well-adapted state is a matter of scientific interest, too, but from an evolutionary point of view it is secondary to the basic fact of adaptation or survival. (Simon, 1983, 37)

In this case we simply say that A is chosen because it is the globally optimal alternative; how it is chosen is not explained. As mentioned in relation to biological functionalism, I agree with Simon (1983, 39) and Elster (1979, 28-35; 1983, ch. 2) that this view is not satisfactory. A full explanation must suggest some mechanism. Otherwise we may misinterpret the facts, as in the case of Alchian's travellers. This problem is much more serious in economics and other social sciences than in biology: in the latter case we have a standard mechanism (natural selection) but this is not so within the social sciences. Therefore, Friedman's 'as-if' theory of adaptation is not satisfactory; the need for a specification of the underlying mechanism is much larger than in the case of biological explanation.

3. Reinforcement-evolutionary explanation (Parijs, 1981, chs. 2, 4; Elster, 1989, ch. 9) shows structural similarity with Lamarckism. It does not assume that agents are able to evaluate A and B without a process of trial-and-error. However, when a favourable behaviour is chosen, it is reinforced because of its revealed comparative advantage (profitability) for the agent. If an inferior alternative is later tried out, the agent returns to his former behaviour which he is able to remember. If we may assume

that the agents have obtained knowledge about A and B, either by direct experience or by obtaining information from other agents, the conclusion is that the relatively persistent selection of A reveals that it is a local optimum under the given circumstances. The localised character of trial-and-error processes does not allow us to judge whether it is a global optimum; maybe an alternative C is better. In a set-up of objective alternatives, a B-user will move towards A but we cannot be sure that alternative C (objectively the global optimum) will be explored.

4. Natural-selection explanation may also be called competitive-evolutionary explanation to avoid the direct invocation of the similarity with a Darwinian account (Parijs, 1981, ch. 3; Elster, 1989, ch. 8; Matthews, 1984/93). Like the reinforcement-evolutionary explanation, it does not assume that agents are able to make *ex ante* evaluations of the alternatives. Some happen to become A-users and others B-users. But once the agents have made their choice, they become inflexible. Supposing that A is the superior alternative, the consequence is that A-users perform better in the competitive struggle than B-users. In the end B-users may be eliminated by bankruptcy or take-overs. Supposing such a process in the case to be explained, we may conclude that A is a local optimum but we cannot say whether it is a global optimum since we cannot assume that the firms have explored the whole space of possible alternatives. If we do not know whether A is a relatively persistent behaviour, we cannot say anything about the optimal characteristics of the choice.

5. Saltationist explanation of evolutionary-economic facts is based on the supposed existence of two types of change (Awan, 1986/91; Mokyr, 1990). As in biology such an explanation was proposed at the beginning of the century (Schumpeter, 1910; 1912/34; Wieser, 1907/29; 1914/27). The basic idea is that the normal process of evolutionary adaptation is ultimately conservative; therefore, the process of evolution will come to an end unless extraordinary evolutionary jumps take place. Since the latter changes are more difficult to implement than adaptive changes, both with respect to entrepreneurial capabilities and with respect to system resources, the evolutionary process shows shifts between periods dominated by the one or the other. After the introduction of a set of major evolutionary novelties, the economic system becomes characterised by a process of adaptive changes which follow a trajectory in much the same way as pointed out by the orthogenetic explanation within biology; only after a period of time is there room for another saltation. We may also talk of the jump between different paradigms and the incremental adaptations within a paradigm (see sections 2.5-2.7). This type of an evolutionary process makes it necessary to explain in a historical context the persistent use of alternative A: if A and B are both well-developed alternatives within an existing techno-economic paradigm, then we may consider A as a local optimum; if A belongs to an old paradigm and B belongs to a potential paradigm, then the evaluation of the dominant

decision-making becomes much more difficult; B may succeed together with the new paradigm but this demands an extraordinary effort with respect to leadership and resources as well as a good deal of luck.

The evolutionary-economic forms of explanation (3, 4, and 5) are clearly very different both with respect to the selection units (e.g., intrafirm decision-making alternatives or the firms themselves), dominance of preservation over variety-creation, emphasis on micro- or macro-evolutionary events, etc. However, they are in principle compatible with the definitions and characteristics presented in sections 1.1 and 1.2.1. In my opinion, they are, rather, designed to treat different classes of potentially evolutionary phenomena. A further discussion of them seems to presuppose a specification of the kind of economic system which is evolving according to a certain selection mechanism.

Evolutionary economics cannot rely on a standard form of explanation to the same extent as evolutionary biology. The toolbox of evolutionary economics must probably include the mechanisms of reinforcement selection as well as of natural selection, and it cannot ignore the more concrete mechanisms suggested by the saltationist approach to economic evolution. Similarly, the 'dialogue' between old and new evolutionary economics cannot rest on a single type of explanation. For example, we should deal with both Marshall's gradualist approach which emphasises (with Darwin) that *Natura non facit saltus*,[4] nature does not make a leap, and Schumpeter's (1909/52, 8; 1912/34, 216) saltationist approach with the motto: *Natura facit saltus*. The difference is to a large extent a matter of emphasis: Marshall was mainly interested in the more or less automatic increase in knowledge and productivity which is the outcome of the gradual increase in scale of production; Schumpeter was interested in mechanisms which allowed the study of the preconditions for and consequences of radical, irreversible, non-repetitive change. A modern evolutionary-economic synthesis has to cover both these aspects of economic evolution.

Notes

1   The shortness of the exposition does not allow a real treatment of the historical forms of biological explanation. An exploration of this subject is found in Mayr (1982).
2   The resistance to the Darwinian answer was not only based in creationism, Lamarckism and saltationism but also in the fact that such a Darwinian-type of evolutionary explanation is not easily reconciled with well-established beliefs about the proper conduct of science. This point has not only been made by Popper (1972; 1974/82); it was already present in reviews of Darwin's work (Hull, 1973) which show that 'many of the reviewers were competent scientists honestly trying to evaluate a novel theory against the commonly accepted standards of scientific excellence, and evolutionary theory consistently came up as wanting.' First of all, Newton's theory had apparently been proved by its ability to suggest predictions of future celestial phenomena while Darwinian theory did not allow precise predictions; instead certain formulations about the survival of the fittest appeared to be (nearly) tautological. Second, Darwin had to rely on hypotheses on heredity and species-formation which in his time appeared undecidable. Third, his theory presupposed a

dismissal of the principle of the immutability of species which to many observers appeared to lead into an intellectual quagmire. These and many other objections led to a rejection of Darwinism, but later developments have shown that they should rather have led to a modification of the meta-scientific rules which were used by the criticism (Hull, 1974; Mayr, 1982).
3   A huge literature exists about this type of explanation. A pedagogical starter is Elster (1989, chs. 3-4)
4   Marshall's (1890/1961, iii, xiii, 249) motto for his classic textbook, '*Natura non facit saltum*', is taken from Darwin (1859/1964, 194, 206, 243, 471) but *saltus* is the correct form rather than *saltum*.. The expression relates to the methodological ideas and formulations of Linné and Leibnitz as well as to older sources.

Appendix:
Algorithmic notation and the programming of Nelson-and-Winter models

A.1. Algorithmic modes of describing evolutionary models

A.1.1. The notation of this book

The restatements of Nelson and Winter's models as well as other parts of evolutionary modelling have been formulated in a notation which includes many aspects of imperative programming languages, not least the ones used in academic communication: ALGOL, SIMULA, PASCAL, MODULA, etc. It has, especially, been attempted to develop a PASCAL-like mode of expression (see Jensen and Wirth, 1974).[1] The reason is, of course, not to present computer programs. Instead it has been attempted 1) to express clearly the issues involved in the development of a computer-implementable concept of evolutionary-economic processes, 2) to sharpen the language used for communicating about evolutionary issues among economists. In both cases I believe that the language of computational processes has been much too slow in its diffusion into the communication among most economists in the areas treated in the present book. This language is quite easy and many economists interested in evolutionary issues have some knowledge of programming.

The linguistic unit in the notation is the statement. Statements are separated by ';'. Within a statement parentheses (and brackets) are used to avoid ambiguities or to increase readability. Indentation is used as a partial alternative to parentheses as well as for increasing readability. There are two major types of statements: definitions and commands.

1) A definition includes a name, the definition sign '$\equiv$', and the body of the definition. In the above text many definitions are made verbally and even implicitly and not by definition statements. However, it is sometimes useful to spell them out.

The conventions about naming of constants, variables and functions are relatively loosely specified. Within definitions of functions there is often a need for variables with no relevance for the overall computation. These are local variables which in principle may be given local names which should not be confused with the names of those of the global variables, which are not used in the context, even if they have the same names. However, the reader may have problems with this naming convention so I try to avoid overlaps within each chapter (instead Greek letters are used).

The specification of user-defined data structures is left for further programming. However, several aspects of the data structures can be read from the notion. A time-varying variable is denoted X_t, while other types of subscripts are used to denote an array of variables: $X_{i,t}$ is indicating the existence of the array $X_t \equiv (X_{1t},\dots,X_{it},\dots,X_{nt})$, where n is the number of elements in the array. Superscripts are used to distinguish different but semantically similar variables: X_t^a and X_t^b and $\overline{X}_t^a$ are thus three different variables. Sometimes helping variables are introduced, e.g. $X_{i,t+1}$ which means that we have an extra array, $X_{t+1} \equiv (X_{1,t+1},\dots,X_{i,t+1},\dots,X_{n,t+1})$, for storing intermediary results. At the end of the computations the results contained in X_{t+1} are transferred to X_t, and we are ready to make the calculations for the next period. This is indicated by '$t := t+1$'.

The notation is built on the assumption that the programmer is allowed to define his own functions and procedures with a suitable number of arguments, e.g. '$F(x,y)$'. Normally the full specification of the user-defined functions and procedures is left to a more detailed level of programming than used in the present book. In principle they should be defined in a definition statement, e.g. '$F(x,y) \equiv z$', where z is a sequence of statements using the arguments/parameters x and y as well global variables, other constants and expressions. When specifications of any of the variables are needed the definition may end by a condition, e.g., ... where $y \in \{\underline{\text{true}}, \underline{\text{false}}\}$.

2) Commands are the most conspicuous aspects of imperative computer languages. They combine expressions and command structures.

An expression is a part of a statement which is supposed to be evaluated by the reader or the computer system in order to provide information for the execution of the statement. The notation is freely using mathematical and statistical expressions. Furthermore, logical expressions, like '$x > y$', are used. In order to evaluate such an expression we first evaluate the two subexpressions 'x' and 'y'"and then the expression as a whole. The value of the evaluation of a logical expression is either <u>true</u> or <u>false</u>. The normal equality sign, '=', is only used to denote equality in logical expressions.

The reason why some programming languages are called imperative is because of the dominant role of the assignment statement. Such a statement replaces the value (the content) of a variable with the outcome of an evaluation. The assignment statement consists of the name of a variable (or more variables), the assignment sign (':=' or '←') and a more or less complex expression.

There are three types of structure controlling the computations: sequential computation is denoted by the succession of statements; iteration is denoted by '<u>while</u> x <u>do</u> y' or '<u>repeat</u> n <u>times</u> y', where x is a logical expression and n is an integer number; selection is denoted '<u>if</u> x <u>then</u> y <u>else</u> z' or '<u>if</u> x <u>then</u> y' (else do nothing).

Another element in the notation is constraints on variables and definitions: '<u>constraint:</u> $x > y$' means that the computer system ensures that expression x is always greater than expression y.

There are, of course, limitations to the kinds of problem that this PASCAL-like notation is suited for. An interesting exercise would be to try to express the search and productivity part of the Nelson-and-Winter model in a functional-style language like LISP or the whole model in one of the object-oriented versions of computer languages, cf. Abelson and Sussman (1985).

The notation should be sharply distinguished from actual programming. For readers interested in simple simulation exercises the DYNAMO/STELLA program found in a footnote to chapter 3 may be implemented and filled out with a variety of parameter values. For others it may be used to check the structure of the calculations behind figure 3.4 (see Goodwin, 1990a, 45). For still others it may just be an example of the computerese which it is still necessary to write, even by means of user-friendly programming tools.

A.1.2. The analytical cycle

The following listing of programs (procedures and data structures) for evolutionary-economic experiments is intended to support most of the steps in the analytical process. When making computer experiments you normally start from an existing model and the related computer program. Then you create suitable data structures for the number of firms you want to study (1). Then you set the parameters of the model (2), 'run' the model (3) and analyse the results (4). Of course, analytical work does not stop here. Normally, you start with a more or less equilibrated state of the model later you study the consequences of changing different parameters (5) and re-run the model and compare the new results with the old ones. You may also want to revise the basic data structures of the simulation model (6).

These steps are well-understood from, e.g., macroeconomic computer simulation (cf., e.g., Marris, 1991). However, the crucial step is much too often missing. Often the most obvious reaction is to revise your basic theory and behavioural model rather than just its parameters. But this is normally impossible because of the sheer complexity of the task and for the practical reason that you have to leave the work with the programmed version of the model, reprogram the model (in, say, PASCAL), recompile the result and hope that the new version will function when you set its parameters and run it. The practical problem has, however, been easy to solve for some time. The reason is that an interactive approach to the reprogramming has been developed, not least by the Artificial Intelligence community who from the very beginning found out that a computer system which can interpret the programming statements continuously, and immediately allow inspection of the results, is well-fitted for creative analytical work. At the same time it is important that the programs (procedures) can be manipulated by (meta)programs which take simple programs as 'data'. This flexibility is well known from the programming language LISP (and the related dialects like SCHEME, cf. Abelson and Sussman, 1985). Some of this flexibility is found in MAPLE, a mathematical manipulation language and the related library of routines. One reason for choosing MAPLE is that this package combines mathematical and modelling power with the interactive approach known from, e.g., Artificial Intelligence experiments (thus the MAPLE programming language is, e.g., interpreted rather than compiled). It is possible to design a programming environment in terms of a meta-package (let us call it EVOLVE). Such a package may in the future include pre-programmed packages of Nelson-and-Winter models, Axelrod models, etc.

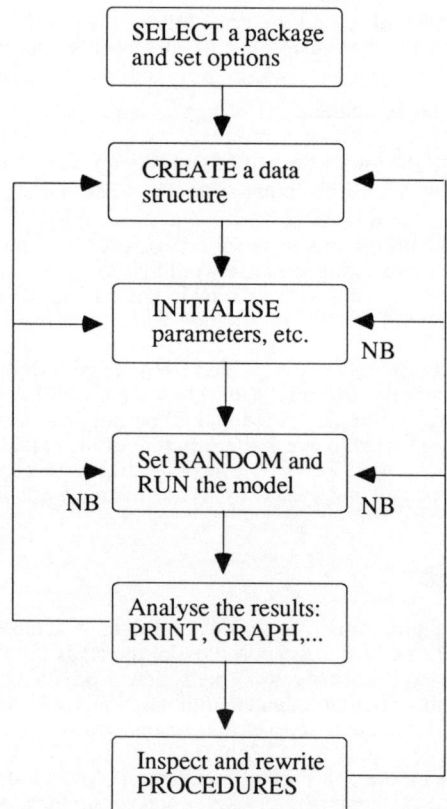

Figure 1. A possible EVOLVE meta-package and the related analytical process.

The structure of such an EVOLVE meta-package is shown in figure 1. The user chooses one model for closer analysis. At the same time he may set certain options. More importantly, the package allows him to inspect and rewrite different parts of the program while the data structure and initialisation values are still in the memory of the computer, and he may also make experiments which simply consist in redefining the parameter of the (pseudo-) random number generator (see the NB's of figure 1). If such activities are repeated, he will see an 'evolving' set of evolutionary-economic packages rather than a (FORTRAN) program which is facing a strong lock-in to a narrow trajectory.

A.1.3. The MAPLE notation

Computer programming may also be used to explore and expand the structure of more complex models like the Nelson-and-Winter family of models and the Axelrod and Trader's Game models. For this purpose we need a programming language which supports a wide variety of mathematical and computational concepts. The problem with these kinds of models is that they are too complex to analyse fully in a formal way. Instead computer simulations are a major tool of analysis. However, the information transferred from the researcher who has invented and analysed a certain model to his readers is rather limited and sometimes even confusing, as any reader of Nelson and Winter's (1982) presentation of their simulation experiments can witness. Winter (1984/91; 1986) goes a long way to remedy these problems. But still he appears to be bound to the original programming (which has not been documented but which is probably like the FORTRAN program presented in the PhD thesis of Schuette, 1980).

To help in rethinking, modifying and expanding the Nelson-and-Winter models, I have tried to develop the skeleton of a software system which allows easy access to the world of evolutionary conceptualisation and modelling. It seems that the MAPLE package, which is based on the work of the Symbolic Computation Group at the University of Waterloo, Canada, is surprisingly clear-cut, both with respect to the computer science foundations of its language and with respect to the openness of the system of procedures which constitutes its major feature (Char *et al.*, 1990, 1991a, 1991b).[2] Thus, even if MAPLE is much less known than another mathematics package, MATHEMATICA, MAPLE seems better suited for developing an evolutionary-economic analysis.

This kind of programming notation may seem quite 'hostile' to many economists. Therefore, it must be emphasised that it represents a rather standardised mode of communication both between researchers and between the researcher and the computer. We are, in other words, dealing with a *lingua franca* which is central to specialised forms of communication just as Latin and English are in more informal affairs. The notation includes many aspects of different programming paradigms (imperative, functional, object-oriented, etc.), but it comes closest to imperative languages like ALGOL, SIMULA and PASCAL. An overview of the MAPLE statements is given in table A.1 (next page)

The linguistic unit in the notation is the statement. Statements are separated by ';'. Within a statement parentheses are used to avoid ambiguities or to increase readability. Indentation is used as a partial alternative to parentheses to increase readability.

The reason why some programming languages are called imperative is because of the dominant role of the assignment statement which is basically a name which points to a value. The first time a name is used on the left side of an assignment statement, the name is placed in the first column of a 'table of variables' (the 'environment'). In another column the value of the variable is placed. Through a later assignment statement this value (the content) of a variable may be replaced by the outcome of an evaluation of an expression. In any case, the assignment statement consists of the name of a variable, the assignment sign (':=' or '←')[3] and a more or less complex expression.

Table A.1. Overview of MAPLE statements[4]

Name	Syntactic Form	Meaning	Comments
assignment	<name> := <expression>;	associates a name with the value of an expression, like profit := sales - costs;	'value' should be understood quite generally, e.g., to include the text of a procedure definition, a matrix, etc.
procedure creation	<name> := proc (<nameseq>) local <nameseq>; <statseq> end;	associates a name (a procname) to a sequence of definitions of names for parameters ('arguments') and local variables and to a sequence of statements which applies them, like double := proc(a) a * a; end;	syntactically, this is just an application of the assignment statement; semantically, we have here a core category; it returns the last evaluation but it may do much more through changing global variables
procedure call	<procname> (<sequence of param. values>);	invokes the procedure definition to be performed using the parameter values specified in the call, like double(25);	the result of the procedure call may be used in many ways, like in the assignment statement, e.g., b := double(25);
expression evaluation	<expression>;	the result of stating an expression is an evaluation of it which returns its value	syntactically, an isolated procedure call is just an example of an expression evaluation
statement sequence	<statementseq> <statement>	the series of statements are performed in the order specified; each statement ends with a ';'	
selection	if <expr> then <statseq> else <statseq> fi; if <expr> then <statseq> elif <expr> then <statseq> fi; etc.	causes the expression to be evaluated (as true or false); if the result is true then the first sequence of statements is performed, otherwise the eventual statseq after else is performed; elif is a simplification	fi should be understood as a closing parenthesis which helps the MAPLE interpreter to define the scope of the if sentence
repetition	for <name> from <expr> to <expr> do <statseq> od;	repeats the execution of the statements for the number of times defined by the expressions and, optionally, while a certain condition is true	od should be understood as a closing parenthesis which helps the MAPLE interpreter to define the scope of the for sentence

read	read <expr>;	causes a file to be read into the present MAPLE session, like read `Macintosh HD: Maple: lib: EVOLVE-lib: Nelwin1: NWdata.m`;	luckily, the correct syntactic form for the pathnames of files can be inserted into a read statement by the facility in the EDIT menu to 'paste' an existing pathname
save	save <expr>;	causes the current environment to be saved under the filename specified	see read

A definition of a procedure begins with a name, the assignment sign ':=', the reserved word 'proc' and a set of parentheses enclosing none or more names of arguments of the procedure, e.g.:

• *RandomNormal := proc(mean, sigma)*

In this case there is a function for generating random numbers with two parameters (mean and sigma) which are unspecified at the moment when the procedure is being defined. The definition is given in the body of the definition which goes from the heading already specified to the reserved word "end". In this body the actions taken by the procedure are specified.

There are three types of structure controlling the computations: sequential computation is denoted by the succession of statements each of which is ended by a semicolon; iteration is denoted by 'for i from j to n do y;', where x is a logical expression and n is an integer number. The condition 'while x do y;' may be *included* in the if statement (see Char *et al.*, 1990a, 41). Selection is denoted 'if x then y else z;' or 'if x then y;' (else do nothing), etc.

Such commands are a conspicuous aspect of computer languages. They combine expressions and command structures. An expression (see table A.2) is a part of a statement which is supposed to be evaluated by the reader or the computer system in order to provide information for the execution of the statement. The notation is freely using mathematical and statistical expressions. Furthermore, logical expressions, like '$x > y$', are used. In order to evaluate such an expression we first evaluate the two subexpressions 'x' and 'y' and then the expression as a whole. The value of the evaluation of a logical expression is either true or false. The normal equality sign, '=', is only used to denote equality in logical expressions.

The conventions about naming of constants, variables and functions are relatively open. Within definitions of procedures there is often a need for variables with no relevance for the overall computation. These are 'local' variables which in principle may be given local names which should not be confused with the names of those of the 'global' variables, even if they have the same names. To avoid confusion, a list of local names is specified at the beginning of the procedure, while the global names are often written with an underline, like _HLEVEL for the variable specifying the level of help wanted by the user of the system.

The names of variables are often characterised by indices for firm number and period number. A time-varying variable is denoted x[t] or x.t (t concatenated to x giving the name 'xt') and a firm-oriented variable x[i]. In combination we have the variable x for the i'th firm and the t'th period like x.t[i] or x[t,i].

These few comments on the notation are far from exhaustive. They may, however, help to emphasise the necessary conditions for the user to cover the whole of the analytical cycle.

Table A.2. Some MAPLE expressions and data structures[5]

Name	Syntactic form	Meaning	Comments
expression sequence (exprseq)	expr1, expr2, expr3, ...	the sequence of expressions ready for storing or use	the sequence may consist of quite different expressions, like sets, matrices, etc.
list	[<exprseq>]	a sequence of expressions within brackets, or a sequence of lists, of lists of lists, of sets, etc., like [a,b, [c, d], e]	a list is best understood by considering the operations for inspecting and changing the list, e.g. list1 := [a,b, [c, d], e]; op(3,list1); --> [c,d] op(3,op(2,list1)); --> d
set	{<exprseq>}	a sequence of expressions within braces, see list	
equation	<expr> = <expr>	two expressions related by an equation sign often denote that the left-hand expression has the property of the right-hand expression, like profit = 27	
equation sequence (eqseq)	equat1, equat2, equat3, ...	the sequence of equations ready for storing or use	
table	table([<eqseq>])	a structured data type, like table([sales = 100, costs = 73])	a table is best understood by considering the operations for inspecting and changing the list, e.g. firm1 := table([sales = 100, costs = 73]); firm1[sales]; --> 100;
range	<expr> .. <expr>	a range for the definition of an array like 1 .. n	
array	array(<rangeseq>, <valuelist>)	a highly structured data type, like array(1..2, [100, 73])	an array is best understood by considering the operations for inspecting and changing the list, e.g. firm1 := array(1..2, [100, 73]); firm1[1]; --> 100;
matrix	matrix(<rowdim>, <columndim>, <valuelist>);	a matrix is a two-dimensional array which allows the operations of linear algebra by means of a special MAPLE package (linalg), an example of a matrix is matrix(2,2,[a,b,c,d])	operations like multiply(A,B); scalarmul(A,2); transpose(A);
selection	<name> [<index>]	this lookup function gives the name of a table or an array followed by an index number	see the example of an array operation

uneval-uated expression	'<expr>'	quotes tells MAPLE to avoid immediate evaluation of an expression	quotes only postpone evaluation; in a two step procedure the expression is evaluated in the second round
string	`<sequence of chars and blanks>`	strings are always left unevaluated	the maximum length is 499 chars and blanks; unfortunately, strings are printed in a non-intelligent way by MAPLE
special constants	true, false, NULL, etc.		you may assign a truth value to a variable like optimism := true;

A.2. Nelson-and-Winter models formulated as MAPLE programs

In the following formulation of functions and procedures, the classic Nelson and Winter model family has been formulated according to the paradigm of structured programming (Dahl *et al.*, 1972). The core of the program (or system of functions and procedures) is the function STATE (section A.2.2) which is formulated in terms of computational recursion. This means that the user makes his orders to the computer in terms of the final result he wants to see, e.g., the state of an industry of 16 firms after 100 periods of economic activity. Since the model is conceived as a Markov process, this result may be obtained if we know the last (99th) state of the process. If not, STATE is called once more. In this way the process goes on until we reach a defined state.

If the user starts a series of experiments, the recursion has to go back to the 'bottom state' in the first run, i.e. the computer finds out that nothing has yet been computed and it therefore has to apply the initialisation procedure for period 0 (see section A.2.3). However, the next time the user requires a computation the recursive STATE function will take advantage of previously computed states.

The parameters used in these computations are mainly provided by an initial invocation of PAR and SEARCHSP—and for some experiments also by BinaryDefPAR (see section A.2.1). However, in order to equilibrate the industry in the initial period, some parameters and the initial capital provision of firms must be specified in a way which is dependent upon the number of firms of the industry. This takes place through InitialSTATE (section A.2.3).

The probabilistic mechanism of transformation from one state to the next is specified by the functions called by the STATE function. ComputeSTATE (section A.2.4) divides the labour of computation between MarketProcess (section A.2.5), NewTechno (section A.2.6) and NewCapital (section A.2.7) which in turn evoke other functions which are listed immediately after the specification of their function at the higher hierarchical level.

A.2.1. Data structures

Parameters

Parameter values are stored by this function and most of them are also initialised here. r_im and r_in are initialised from the initialiseSTATE function since they are n-dependent and we do not know n before we come to this function. The user can change any parameter value. However, some parameters must be changed by 'meta-parameters'. The initialisation values follow Nelson and Winter, 1982, especially ch. 12.

• PAR := proc()

option remember;

```
if nargs = 0 then

    PAR( A_init )      :=   0.16      ;
    PAR( b )           :=   2.5       ;
    PAR( c )           :=   0.16      ;
    PAR( delta )       :=   0.03      ;
    PAR( dem_shift )   :=   0         ;
    PAR( d_im )        :=   1.25      ;
    PAR( d_in )        :=   0.125     ;
    PAR( eta )         :=   1         ;
    PAR( imi_share )   :=   0.012     ;
    PAR( inno_share )  :=   0.12      ;
    PAR( K_min )       :=   0         ;
    PAR( phi )         :=   0.01      ;
    PAR( P_max )       :=   10        ;
    PAR( psi )         :=   1         ;
    PAR( sigma_in )    :=   0.05      ;
    PAR( tot_rvn )     :=   67        ;

    PAR( innoSplit )   :=   true         ;
    PAR( markUp )      :=   Winter_84    ;
    PAR( technoType )  :=   scienceBased ;
    PAR( binaryPAR )   :=   false        ;
    PAR( binaryDef )   :=   0, 0, 0, 0   ;

    PAR( digits )      :=   6         ;
    PAR( digits_out )  :=   3         ;
    PAR( rand_seed )   :=   1         ;

    readlib(forget);
    forget(PAR,``);
    forget(STATS);

fi;

end;
```

A binary mode of defining parameters

As in chapter 13 of Nelson and Winter, 1982, four parameters are defined. High is '1' and low is '0'. For example, BinaryDefPAR(0, 1, 1, 1); parameter values are from Nelson and Winter, 1978, pp. 546 f. but some rounding makes, e.g., the 4-period increase in the latent technique a little more than 2% and 6%.

• BinaryDefPAR := proc()

local i;

    PAR(tot_rvn) := 64;

```
for i from 1 to 4 do
    if not member(args[i], {0, 1}) then
        ERROR(
            `PAR(binaryDef) must be defined as a sequence of`.
            `four binary numbers (either 0 or 1), e.g. `.
            `PAR(BinaryDef) := 0, 1, 1, 1;`
            );
    fi;
od;

if args[1] = 1 then
        PAR( eta )        := 1000;
else
        PAR( eta )        := 1;
fi;

if args[2] = 1 then
        PAR( d_im )       := 0.005;
else
        PAR( d_im )       := 0.0025;
fi;

if args[3] = 1 then
        PAR( phi )        := 0.015;
    if args[4] = 1 then
        PAR( sigma_in )   := 0.18;
    else
        PAR( sigma_in )   := 0.06;
    fi;
else
        PAR( phi )        := 0.005;
    if args[4] = 1 then
        PAR( sigma_in )   := 0.06;
    else
        PAR( sigma_in )   := 0.02;
    fi;
fi;

end;
```

The search space

Values of points in the search space are stored by this function. The values are generated randomly within an interval (here: 0..2). The initial value of A, PAR(A_init), is inserted into the space. At any time, the search space can be given new random values, e.g., by SEARCHSP(60, 0..4); The individual points in the search space can be manipulated by SEARCHSP(13) := 0.45; — e.g. in order to create larger jumps.

• SEARCHSP := proc()

local A_list, A_space, i, Possibility;

option remember;

   if nargs = 0 then

      SEARCHSP(1) := 0.05 ;
      SEARCHSP(2) := 0.06 ;
      SEARCHSP(3) := 0.11 ;
      SEARCHSP(4) := 0.16 ;
      SEARCHSP(5) := 0.21 ;
      SEARCHSP(6) := 0.29 ;
      SEARCHSP(7) := 0.30 ;
      SEARCHSP(8) := 0.36 ;
      SEARCHSP(9) := 0.38 ;
      SEARCHSP(10) := 0.40 ;
      SEARCHSP(11) := 0.43 ;
      SEARCHSP(12) := 0.44 ;
      SEARCHSP(13) := 0.56 ;
      SEARCHSP(14) := 0.68 ;
      SEARCHSP(15) := 0.72 ;
      SEARCHSP(16) := 0.74 ;
      SEARCHSP(17) := 0.84 ;
      SEARCHSP(18) := 0.85 ;
      SEARCHSP(19) := 0.92 ;
      SEARCHSP(20) := 0.98 ;
      SEARCHSP(21) := 1.06 ;
      SEARCHSP(22) := 1.07 ;
      SEARCHSP(23) := 1.13 ;
      SEARCHSP(24) := 1.16 ;
      SEARCHSP(25) := 1.18 ;
      SEARCHSP(26) := 1.20 ;
      SEARCHSP(27) := 1.32 ;
      SEARCHSP(28) := 1.33 ;
      SEARCHSP(29) := 1.42 ;
      SEARCHSP(30) := 1.48 ;
      SEARCHSP(31) := 1.50 ;
      SEARCHSP(32) := 1.52 ;
      SEARCHSP(33) := 1.67 ;
      SEARCHSP(34) := 1.77 ;
      SEARCHSP(35) := 1.80 ;
      SEARCHSP(36) := 1.83 ;
      SEARCHSP(37) := 1.90 ;
      SEARCHSP(38) := 1.91 ;
      SEARCHSP(39) := 1.93 ;
      SEARCHSP(40) := 1.96 ;

      SEARCHSP(latent, 0) := PAR(A_init);

   elif nargs = 2 then

```
        if not
            (type(args[1], integer) or type(args[2], integer)) then
            ERROR(
                `the number of points in the search space and the`,
                ` highest possible productivity must be integers.`
                );
        else
            forget(SEARCHSP);
            Digits := 3;
            Possibility := MakeRandUniformDistr(0..args[2]);
            A_set :=    {PAR(A_init)} union
                        {seq(Possibility(), i = 1..args[1])};
            A_list := convert(A_set, list);
            A_space := sort(A_list, numeric);
            for i from 1 to args[1] do
                SEARCHSP(i) := A_space[i];
            od;
        fi;
    fi;

end
```

A.2.2. The recursive master procedure

This recursive function is reflecting the Markov-chain logic and is at the hierarchical top of StructuredNelwin modelling. It places most of the other functions in the hierarchy through direct and indirect connections. It allows the user to expand the number of periods under study without a total and time-consuming recalculation of the previous results.

• STATE := proc(n, t)

option remember;

```
    if t = 0 then

        STATE(n, t) := InitialSTATE(n);
        _seed := PAR(rand_seed);

    else

        STATE(n, t) := ComputeSTATE(n, STATE(n, t - 1), t);

    fi;

end;
```

A.2.3. State initialisation

Capital is initialised in a way which puts the industry into equilibrium at the prevailing productivity levels. Innovative and imitative search costs per unit of capital are initialised so that total search costs will be the same for cases with different numbers of firms in the industry, cf. Nelson and Winter, 1982, p. 302. Nelson and Winter use different formulations of the target mark-up factor in their works. To make it easy to change, a special function for its computation is formulated. TargetMarkUp(marketShare) is defined after the DesiredInvest function below.

• InitialSTATE := proc(n)

local i, initSTATE, marketShare, targetMarkUp, tot_imi_costs, tot_inno_costs;

```
    if not type(n/2, integer) then
        ERROR(`the number of firms must be even`);
    fi;
    PAR(N_firms) := n;
    Digits := PAR(digits);

    initSTATE := array(1..2, 1..n);

    for i from 1 to n do
        initSTATE[1,i] := SEARCHSP(latent, 0);

        marketShare := 1/n;
        targetMarkUp := TargetMarkUp(marketShare);
        initSTATE[2,i] := PAR(tot_rvn)/
                          (n*PAR(c)*targetMarkUp);
    od;

    tot_inno_costs := PAR(tot_rvn)*PAR(inno_share)/2;
    tot_imi_costs := PAR(tot_rvn)*PAR(imi_share);
    for i from 1 to n do
        if i <= n/2 then
            PAR(r_in, i) :=
                (tot_inno_costs/(n/2))/initSTATE[2,i];
        else
            PAR(r_in, i) := 0;
        fi;
        PAR(r_im, i) := (tot_imi_costs/n)/initSTATE[2,i];
    od;

    STATS(A_max, 0) := SEARCHSP(latent, 0);
    MarketProcess(n, initSTATE, 0);

    RETURN(op(1, initSTATE));

end;
```

A.2.4. *The general computation of the state variables*

The task of computing the next state of the Markov chain by means only of the present state is here split into subtasks: 1) the short-term market process (MarketProcess), 2) the search for and choice of technique (NewTechno) and 3) the depreciation of existing physical capital and the creation of new capital.

• ComputeSTATE := proc(n, STATEinfo, t)

local i, nextSTATE, P, PandTQ, presentSTATE, TQ;

    Digits := PAR(digits);

    presentSTATE := STATEinfo;

    PandTQ := MarketProcess(n, presentSTATE, t);
    P := PandTQ[1];
    TQ := PandTQ[2];

    A_max := max(seq(presentSTATE[1,i], i = 1..n));
    STATS('A_max', t) := A_max;

    nextSTATE := array(1..2, 1..n);

    for i from 1 to n do
        if presentSTATE[1,i] = 0 then
            nextSTATE[1,i] := 0;
        else
            nextSTATE[1,i] := NewTechno(
                          presentSTATE[1,i],
                          A_max,
                          presentSTATE[2,i],
                          t
                          );
        fi;
    od;

    for i from 1 to n do
        if presentSTATE[2,i] = 0 then
            nextSTATE[2,i] := 0;
        else
            nextSTATE[2,i] := NewCapital(
                          presentSTATE[1,i],
                          nextSTATE[1,i],
                          presentSTATE[2,i],
                          P,
                          TQ
                          );
        fi;
    od;

    RETURN(op(1, nextSTATE));

end;

A.2.5. The short-term market process

• MarketProcess := proc(n, presentSTATE, t)

local i, price, supply;

    supply := sum(
                presentSTATE[1,i]*presentSTATE[2,i],
                i = 1..n
                );

    STATS(TQ, t) := supply;

    if supply < PAR(dem_shift) then
        price := PAR(P_max);
    else
        price := PAR(tot_rvn)/supply;
    fi;

    STATS(P, t) := price;

    RETURN([price, supply]);

end;

A.2.6. Finding the new technology of a firm

Two alternative regimes of technical change are specified. The "science based" technology case is used in Nelson and Winter, 1982, chs 12-14, while the "cumulative" technology case is used in chs. 14 and 9.

• NewTechno := proc(A, A_max, K, t)

local A_im, A_in, A_latent, A_next;

    if K = 0 then
        RETURN(0);
    fi;

    if PAR(technoType) = scienceBased then

        SEARCHSP(latent, t) := SEARCHSP(latent, t-1)*(1 + PAR(phi));
        A_in := Innovate(SEARCHSP(latent, t), K);

    elif PAR(technoType) = cumulative then

        A_in := Innovate(A, K);

    fi;

    A_im := Imitate(A_max, K);

```
    A_next := max(A, A_in, A_im);

    RETURN(A_next);

end;
```

The firm's innovation process

```
• Innovate := proc(A_mean, K)

local A_in, A_prelim, j, Lottery, NormalDistr, Poisson_mean;

    Poisson_mean := PAR(d_in)*PAR(r_in)*K;
    Lottery := MakeRandPoissonDistr(Poisson_mean);
    NormalDistr := MakeRandNormalDistr(A_mean, PAR(sigma_in));

    if Lottery() > 0 then
        A_prelim := NormalDistr();
        j := 1;
        while SEARCHSP(j) < A_prelim do
            j := j + 1;
        od;
        A_in := SEARCHSP(j);
    else
        A_in := 0;
    fi;

    RETURN(A_in);

end;
```

The firm's imitation process

```
• Imitate := proc(A_max, K)

local A_im, A_prelim, j, Lottery, Poisson_mean;

    Poisson_mean := PAR(d_in)*PAR(r_im)*K;
    Lottery := MakeRandPoissonDistr(Poisson_mean);

    if Lottery() > 0 then
        A_im := A_max;
    else
        A_im := 0;
    fi;

    RETURN(A_im);

end;
```

A.2.7. The firm's investment and depreciation process

• NewCapital := proc(A, A_next, K, P, TQ)

local I_act, I_des, I_max, K_next;

    I_des := DesiredInvest(A, A_next, K, P, TQ);
    I_max := FinanceInvest(A, K, P);

    I_act := max(0, min(I_des, I_max));
    K_next := K*(1 - PAR(delta)) + I_act;

    if IsBankrupt(K_next) = true then
        K_next := 0;
    fi;

    RETURN(K_next);

end;

The firm's desired investment

The 'price lid' is removed by a high value of PAR(P_max). It stems from Winter, 1984/91; 1986, not from Nelson and Winter, 1982. The top-down formulations are, hopefully, clarifying. However, they necessitate a renewed evaluation of desInvest at the end of the function: 'eval(desInvest)'.

• DesiredInvest := proc(A, A_next, K, P, TQ)

local actualMarkUp, desInvest, deprCapital, desCapital,
desMarkUp, expandContract, marketShare, priceLid, targetMarkUp;

    desInvest := desCapital - deprCapital;

    desCapital := K*(1 + expandContract);
    deprCapital := K*(1 - PAR(delta));

    expandContract := 1 - desMarkUp/actualMarkUp;

    desMarkUp := min(targetMarkUp, priceLid);
    actualMarkUp := P/(PAR(c)/A_next);

    targetMarkUp := TargetMarkUp(marketShare);
    marketShare := A*K/TQ;

    priceLid := .999*(P + PAR(P_max))*A_next/(2*PAR(c));

    RETURN(eval(desInvest));

end;

The target mark-up factor of the firm

The first formulation (Winter_84) of the mark-up function is from Winter (1984/91; 1986) and reduces to the Nelson and Winter (1982) version with appropriate parameter values. The second formulation (NandW_78) is from Nelson and Winter, 1978.

• TargetMarkUp := proc(marketShare)

local targetMarkUp;

    if PAR(MarkUp) = Winter_84 then
        targetMarkUp := (PAR(eta) + (1 - marketShare)*PAR(psi))/
            (PAR(eta) + (1 - marketShare)*PAR(psi) - marketShare);
    elif PAR(MarkUp) = NandW_78 then
        targetMarkUp := PAR(eta)/(PAR(eta) - marketShare);
    fi;

    RETURN(targetMarkUp);

end;

The finance of a firm's investment

• FinanceInvest := proc(A, K, P)

local externalFinance, profit;

    profit := P*A*K - (PAR(c) + PAR(r_in) + PAR(r_im))*K;

    if profit <= 0 then
        externalFinance := 0;
    else
        externalFinance := PAR(b)*profit;
    fi;

    I_max := PAR(delta) + profit + externalFinance;

    RETURN(I_max);

end;

Testing for bankruptcy

• IsBankrupt := proc(K)

local test;

    if K < PAR(K_min) then
        test := true;
    else
        test := false;
    fi;

RETURN(test);

end;

A.2.8. Collection of statistical data

This function remembers different statistical information on the evolving industry. Some of the arguments and the values of the function are assigned by other functions.

• STATS := proc(n, t)

local capConc, i, j, productivMean, totCap;

option remember;

    if nargs = 2 then

        Digits := PAR(digits);

        for j from 0 to t do

            totCap := sum(STATE(n, j)[2,i], i = 1..n);
            STATS(A_ave, j) := STATS(TQ, j)/totCap;
            capConc := sum((STATE(n, j)[2,i]/totCap)^2, i = 1..n);
            STATS(TK, j) := totCap;
            STATS(HF_equiv, j) := 1/capConc;

            productivMean := sum(STATE(n, j)[1,i], i = 1..n)/n;
            productivStdDev :=
                (sum((STATE(n, j)[1,i] - productivMean)^2,
                              i = 1..n)/n)^1/2;
            STATS(A_mean, j) := productivMean;
            STATS(A_stdDev, j) := productivStdDev;

        od;

        forget(STATS, n, t);

    fi;

end;

A.3. Starting an object-oriented version of Nelson-and-Winter models

An object-oriented approach represents a rather fundamental rethinking and reprogramming of the Nelson and Winter family of models in terms of a new programming paradigm. This paradigm has its roots back in the SIMULA programming language and is now spreading both by means of specialised languages like SMALLTALK and as an extension of existing languages like OBJECTPASCAL, OBJECTIVE-C, C+ +,

SCHEME, etc. MAPLE's programming language is not constructed for implementing the fundamental object-oriented concepts like encapsulation of data, messages and classes. For modelling purposes these design principles are, however, implementable in a MAPLE programming environment.

In the present context, object-oriented programming can, with some right, be translated to actor-oriented programming. This means, that the units of the program are actors who react upon key words. These 'Requests' are names starting with a capital letter, like 'Supply', 'InvestmentDemand', 'Decide', etc. The requests may be supplemented by one or more arguments of importance for the response to the request. These arguments are names without an initial capital letter (except in the case of the standard names for the state variables: A and K).

All actors have to be created explicitly. The initial troupe of actors is created at the start of the computation but later new actors (firms) may emerge as the result of entrepreneurial activity. In order to create an actor, the creating procedure must be supplied with a sequence of parameter values which determine much of the behaviour of the actor. Later some actors may, however, change their behavioural pattern because of their decisions to change the rules underlying their behaviour.

• CreateFirm := proc()

local Firm;

    Firm := subs(
        {args},
        proc(Request)
        local A, actualMarkUp, deprCapital, desCapital, desFinance,
            desMarkUp, expandContract, externalFinance, I_act, I_des,
            I-max, imiEffort, info, innoEffort, K, Lottery,
            marketShare, price, priceLid, profit, supply, targetMarkUp;

        A := proc()
        option remember;
        end;
        K := proc()
        option remember;
        end;

       # InitialiseState
        if Request = InitialiseState then
            A() := A_init;
            K() := K_init;

       # Supply (and change technology)
        elif Request = Supply then
            supply := A()*K();
            innoEffort := d_in*r_in*K();
            A(in) := TechnoSpace(Decide, innoEffort, A());
            imiEffort := d_im*r_im*K();
            A(im) := max(
                    seq(
                        Firm.i(TechnoInfo, country, imiEffort),

```
                        i = 1..n
                    )
                );
            A(next) := max(A(), A(in), A(im));
            RETURN(supply);

    # InvestmentDemand
        elif Request = InvestmentDemand then
            totalSupply := args[2];
            price := args[3];
            profit := price*A()*K() - (c + r_in + r_im)*K();
            marketShare := A()*K()/totalSupply;
            targetMarkUp := eta + (1 - marketShare)*psi/
                (eta + (1 - marketShare)*psi - marketShare);
            priceLid := .999*(price + P_max)*A(next)/(2*c);
            desMarkUp := min(targetMarkUp, priceLid);
            actualMarkUp := price/(c/A(next));
            expandContract := 1 - desMarkUp/actualMarkUp;
            desCapital := K()*(1 + expandContract);
            deprCapital := K()*(1 - delta);
            I_des := desCapital - deprCapital;
            desFinance := I_des - profit - delta;
            externalFinance :=
                FinancialSector(Decide, desFinance, profit, K());
            if externalFinance = bankruptcy then
                A() := 0;
                K() := 0;
            else
                I_max := delta + profit + externalFinance;
                I_act := max(0, min(I_des, I_max));
                K() := K()*(1 - delta) + I_act;
                A() := A(next);
            fi;

    # OfficialStats
        elif Request = OfficialStats then
            RETURN(A(), K());
        fi;
    end);

    Firm(InitialiseState);

    RETURN(op(Firm));

end;
```

Notes

1. However, the 'syntactic sugar' of the present notation differs much from PASCAL and no attempt whatsoever has been made to impose the paradigm of 'structured programming' (Dahl *et al.*, 1972) upon the notation. In this respect the style of presentation is closer to the LISP tradition (see Abelson and Sussman, 1985).
2. It was Jørgen Østergaard who pointed towards the fact that the MAPLE mathematics package is well suited for evolutionary modelling (which he tries to develop in a way which includes the analysis of product innovation in his PhD work).
3. In some programming languages like FORTRAN, BASIC and C, the assignment is designated by an equation sign ('=') and this may give rise to some confusion in a context of mathematical analysis.
4. Cf. Char *et al.*, 1991a, ch. 3. My version involves certain simplifications, etc., but they are accepted by the MAPLE interpreter program.
5. Much more about expressions and data structures is found in Char *et al.*, 1991a, as well as in computer-science textbooks.

References

Abelson, H. and Sussman, G.J. (1985), *Structure and Interpretation of Computer Programs*, McGraw-Hill, Cambridge, Mass. and London.
Abernathy, W.J. (1978), *The Productivity Dilemma: Roadblock to Innovation in the Automobile Industry*, Johns Hopkins University Press, Baltimore and London.
___ and Clark, K.B. (1985), 'Innovation: Mapping the Winds of Creative Destruction', *Research Policy*, Vol. 14, 3-22.
___ and Townsend, P.L. (1975), 'Technology, Productivity and Process Change', *Technological Forecasting and Social Change*, Vol. 7, 379-396.
___ and Utterback, J.M. (1975), 'A Dynamic Model of Process and Product Innovation', *Omega*, Vol. 3, 639-656.
___ and ___ (1978), 'Patterns of Industrial Innovation', *Technology Review*, 2-9.
Akerlof, G. (1970), 'The Market for Lemons: Quality Uncertainty and the Market Mechanism', *Quarterly Journal of Economics*, Vol. 89, 488-500.
Alchian, A.A. (1950/93), 'Uncertainty, Evolution and Economic Theory', in Witt (1993), 65-75.
___ (1953), 'Biological Analogies in the Theory of the Firm: Comment', *American Economic Review*, Vol. 43, 600-603.
Allen, P.M. (1981), 'The Evolutionary Paradigm of Dissipative Structures', in Jantsch, E. (ed.), *The Evolutionary Vision: Toward a Unifying Paradigm of Physical, Biological, and Sociocultural Evolution*, Westview, Boulder, Colo., 1981, 25-72.
___ (1988), 'Evolution, Innovation and Economics', in Dosi et al. (1988), 95-119.
Allen, R.L. (1991), *Opening Doors: The Life and Work of Joseph Schumpeter*, 2 vols., Transaction, New Brunswick, N.J. and London.
Andersen, E.S. (1991a), 'Techno-Economic Paradigms as Typical Interfaces between Producers and Users', *Journal of Evolutionary Economics*, Vol. 1, 119-144.
___ (1991b), Schumpeter's Vienna and the Schools of Thought, IKE småskrift 70, Institute for Production, University of Aalborg.
___ (1991c), The Core of Schumpeter's Work, IKE Småskrift 68, Institute for Production, University of Aalborg.
___ (1991d), Reconstructing Theory-Evolution, with Special Respect to Schumpeter, IKE småskrift 69, Institute for Production, University of Aalborg.
___ (1991e), Statics and Development: A First Approximation to Schumpeter's Evolutionary Vision, IKE småskrift 71, Institute for Production, University of Aalborg.
___ (1992a), 'Approaching National Systems of Innovation from the Production and Linkage Structure', in Lundvall (1992), 68-92.
___ (1992b), The Difficult Jump From Walrasian to Schumpeterian Analysis (Paper for the Kyoto Conference of the International Schumpeter Society), IKE Småskrift 78, Institute for Production, University of Aalborg.
___ (1993a), Railroadization as Schumpeter's Standard Case, Department of Business Studies, University of Aalborg.
___ (1993b), 'Review of R. Swedberg 'Schumpeter: A Biography'', to be published in *Journal of Economic Literature*, Vol. 31: December.
___ (1993c), The Evolution of Credence Goods: An Approach to Product Specification and Quality Control, Department of Business Studies, University of Aalborg.
___ (1993d), Schumpeter's Core Concepts on Routine and Innovation, Department of Business Studies, University of Aalborg.

____ and Lundvall, B.-Å. (1988), 'Small National Systems of Innovation Facing Technological Revolutions: An Analytical Framework', in Freeman, C. and Lundvall, B.-Å. (eds.) (1988), *Small Countries Facing the Technological Revolution*, Pinter, London and New York, 1-31.
Aoki, M. (1986), 'Horizontal vs. Vertical Information Structures of the Firm', *American Economic Review*, Vol. 76, 971-983.
Arrow, K.J. (1962/83), 'The Economic Implications of Learning by Doing', in Arrow, K.J., *Collected Papers*, Vol. 5, Belknap Press of Harvard University Press, Cambridge, Mass. and London, 1983-85, 157-180.
____ (1973/83), 'Information and Economic Behaviour', in Arrow, K.J., *Collected Papers*, Vol. 4, Belknap Press of Harvard University Press, Cambridge, Mass. and London, 1983-85, 136-152.
____ (1974), *The Limits of Organization*, W. W. Norton & Company, New York.
Arthur, W.B. (1983), Competing Technologies and Lock In by Historical Events, Working Paper 83-90, International Institute for Applied Systems Analysis, Laxenburg.
____ (1988), 'Competing Technologies: An Overview', in Dosi *et al.* (1988), 590-607.
____ (1991), 'Designing Economic Agents that Act Like Human Agents: A Behavioural Approach to Bounded Rationality', *American Economic Review. Papers and Proceedings*, Vol. 81, 353-359.
____ (1993), 'On Designing Economic Agents that Behave Like Human Agents', *Journal of Evolutionary Economics*, Vol. 3, 1-22.
____, Ermoliev, Y.M. and Kaniovski, Y.M. (1987/93), 'Path-Dependent Processes and the Emergence of Macro-Structure', in Witt (1993), 257-266.
Aumann, R.J. (1987), 'Game Theory', in Eatwell *et al.* (1987), Vol. 2, 460-482.
Awan, A.A. (1986/91), 'Marshallian and Schumpeterian Theories of Economic Evolution: Gradualism versus Punctualism', in Wood (1991), Vol. 4, 436-453.
Axelrod, R. (1980a), 'Effective Choice in the Prisoner's Dilemma', *Journal of Conflict Resolution*, Vol. 24, 3-25.
____ (1980b), 'More Effective Choice in the Prisoner's Dilemma', *Journal of Conflict Resolution*, Vol. 24, 379-403.
____ (1984), *The Evolution of Cooperation*, Basic Books, New York.
____ (1987), 'The Evolution of Strategies in the Iterated Prisoner's Dilemma', in Davies (1987), 32-41.
____ and Dion, D. (1988), 'The Further Evolution of Cooperation', *Science*, Vol. 242, 1385-1390.
Bagwell, P.S. (1974), *The Transport Revolution from 1770*, B.T. Batsford, London.
Bain, J.S. (1956), *Barriers to New Competition*, Harvard University Press, Cambridge, Mass.
Batten, D.F. (1987), 'The Balanced Path of Economic Development: A Fable for Growth Merchants', in Batten, D., Casti, J. and Johansson, B. (eds.), *Economic Evolution and Structural Adjustment*, Springer, Berlin, 1987, 64-85.
Baumol, W.J. (1951/70), *Economic Dynamics: An Introduction*, 3rd edn., Macmillan, New York, 1970.
Berg, M. (1980), *The Machinery Question and the Making of Political Economy 1815-1848*, Cambridge University Press, Cambridge.
Blaug, M. (1962/85), *Economic Theory in Retrospect*, 4th edn., Cambridge University Press, Cambridge, 1985.
Booker, L.B., Goldberg, D.E. and Holland, J.H. (1990), 'Classifier Systems and Genetic Algorithms', in Carbonell (ed.) (1990), *Machine Learning: Paradigms and Methods*, MIT Press, Cambridge, Mass. and London, 235-282.
Bottomore, T. (1992), *Between Marginalism and Marxism: The Economic Sociology of J. A. Schumpeter*, Harvester Wheatsheaf, New York.
Boulding, K.E. (1978), *Ecodynamics: A New Theory of Societal Evolution*, Sage, Beverly Hills, Cal. and London.
____ (1981), *Evolutionary Economics*, Sage, Beverly Hills, Cal. and London.
Bowler, P.J. (1983), *The Eclipse of Darwinism: Anti-Darwinian Evolution Theories in the Decades around 1900*, Johns Hopkins University Press, Baltimore and London.

Boyd, R. and Richerson, P.J. (1985), *Culture and the Evolutionary Process*, University of Chicago Press, Chicago and London.
Brookfield, H. (1975), *Interdependent Development*, Methuen & Co., London.
Casson, M. (1982), *The Entrepreneur: An Economic Theory*, Martin Robertson, Oxford.
Cavalli-Sforza, L.L. and Feldman, M.V. (1981), *Cultural Transmission and Evolution: A Quantitative Approach*, Princeton University Press, Princeton, N.J.
Chamberlin, E.H. (1933/62), *The Theory of Monopolistic Competition: A Re-orientation of the Theory of Value*, Harvard University Press, Cambridge, Mass. and London, 1962.
___ (1953), 'The Product as an Economic Variable', *Quarterly Journal of Economics*, Vol. 57, 1-29.
Chandler, A.D. (1977), *The Visible Hand: The Managerial Revolution in American Business*, Harvard University Press, Cambridge, Mass.
Char, B.W., Geddes, K.O., Gonnet, G.H., Monagan, M.B. and Watt, S.M. (1990), *First Leaves: A Tutorial Introduction to Maple*, 3rd edn., Maple, Waterloo, Can.
___ et al. (1991a), *Maple V Language Reference Manual*, Springer, New York.
___ et al.(1991b), *Maple V Library Reference Manual*, Springer, New York.
Clark, K.B. (1985), 'The Interaction of Design Hierarchies and Market Concepts in Technological Evolution', *Research Policy*, Vol. 14, 235-251.
Clark, N. and Juma, C. (1987), *Long-Run Economics: An Evolutionary Approach to Economic Growth*, Pinter, London and New York.
___ and ___ (1988), 'Evolutionary Theories in Economic Thought', in Dosi *et al.* (1988), 197-218.
Clemence, R.V. and Doody, F.S. (1950/66), *The Schumpeterian System*, reprint edn., Augustus M. Kelley, New York, 1966.
Coombs, R., Saviotti, P. and Walsh, V. (1987), *Economics and Technological Change*, Macmillan, Basingstoke and London.
Cyert, R.M. and March, J.G. (1963), *A Behavioural Theory of The Firm*, Prentice-Hall, Englewood Cliffs, N.J.
Dahl, O.-J., Dijkstra, E.W. and Hoare, C.A.R. (eds.) (1972), *Structured Programming*, Academic Press, London and New York.
Dahmén, E. (1950/70), *Entrepreneurial Activity and the Development of Swedish Industry 1919-1939*, American Economic Association Translation Series, Homewood, 1970.
___ (1988), '"Development Blocks" in Industrial Economics', *Scandinavian Economic History Review*, Vol. 36, 3-14.
Darwin, C. (1859/1964), *On the Origin of Species by Means of Natural Selection*, reprint of 1st edn., Harvard University Press, Cambridge, Mass. and London, 1964.
Dasgupta, P. and Stiglitz, J. (1980), 'Industrial Structure and the Nature of Innovative Activity', *Economic Journal*, Vol. 90, 266-293.
David, P.A. (1985/93), 'Clio and the Economics of QWERTY', in Witt (1993), 267-272.
___ (1993), 'Path-Dependence and Predictability in Dynamic Systems with Local Network Externalities: A Paradigm for Historical Economics', in Foray and Freeman (1993), 208-231.
Davis, L. (ed.) (1987), *Genetic Algorithms and Simulated Annealing*, Pitman, London.
Day, R.H. and Chen, P. (eds.) (1993), *Nonlinear Dynamics and Evolutionary Economics*, Oxford University Press, New York and Oxford.
___ and Eliasson, G. (eds.) (1986), *The Dynamics of Market Economies*, North-Holland, Amsterdam.
___ and Groves, T. (eds.) (1975), *Adaptive Economic Models*, Academic Press, New York.
Dosi, G. (1982), 'Technological Paradigms and Technological Trajectories: A Suggested Interpretation of the Determinants and Directions of Technical Change', *Research Policy*, Vol. 11, 147-162.
___ (1984), *Technical Change and Industrial Transformation: The Theory and an Application to the Semiconductor Industry*, Macmillan, London and Basingstoke.

___ (1988), 'Sources, Procedures and Microeconomic Effects of Innovation', *Journal of Economic Literature*, Vol. 26, 1120-1171.
___ (1990), 'Economic Change and Its Interpretation, or, Is There a "Schumpeterian Approach"?', in Heertje and Perlman (1990), 335-341.
___ (1991), 'Some Thoughts on the Promises, Challenges and Dangers of an "Evolutionary Perspective" in Economics', *Journal of Evolutionary Economics*, Vol. 1, 5-7.
___, Freeman, C., Nelson, R.R., Silverberg, G. and Soete, L. (eds.) (1988), *Technical Change and Economic Theory*, Pinter, London.
Dreyfus, H.L. (1972/79), *What Computer's Can't Do: The Limits of Artificial Intelligence*, 2nd edn., Harper & Row, New York, 1979.
Earl, P.E. (ed.) (1988), *Behavioural Economics*, 2 vols., Elgar, Aldershot.
Eatwell, J., Milgate, M. and Newman, P. (eds.) (1987), *The New Palgrave: A Dictionary of Economics*, 4 vols., Macmillan, London and Basingstoke.
Egidi, M. and Marris, R. (eds.) (1992), *Economics, Bounded Rationality and the Cognitive Revolution*, Elgar, Aldershot.
Ekelund, R.B. and Hébert, R.F. (1975/90), *History of Economic Theory and Method*, 3rd edn., McGraw-Hill, New York.
Eldredge, N. (1989), *Macroevolutionary Dynamics: Species, Niches, and Adaptive Peaks*, McGraw-Hill, New York.
Elster, J. (1979), *Ulysses and the Sirens: Studies in Rationality and Irrationality*, Cambridge University Press, Cambridge.
___ (1983), *Explaining Technical Change*, Cambridge University Press, Cambridge.
___ (1989), *Nuts and Bolts for the Social Sciences*, Cambridge University Press, Cambridge.
Emmeche, C. (1991), *Det levende spil: Biologisk form og kunstigt liv*, Munksgaard, Copenhagen.
Flood, M. (1958), 'Some Experimental Games', *Management Science*, Vol. 5, 5-26.
Flores, F. and Winograd, T. (1986), *Understanding Computers and Cognition: A New Foundation for Design*, Ablex, Norwood, N.J.
Foray, D. and Freeman, C. (eds.) (1993), *Technology and the Wealth of Nations: The Dynamics of Constructed Advantage*, Pinter, London and New York.
Forrester, J.W. (1971), *World Dynamics*, Wright-Allen, Cambridge, Mass.
Foster, J. (1987), *Evolutionary Macroeconomics*, Allen & Unwin, London.
___ (1991), 'The Institutionalist (Evolutionary) School', in Mair, D. and Miller, A.G. (eds.), *A Modern Guide to Economic Thought: An Introduction to Comparative Schools of Thought in Economics*, Elgar, Aldershot, 1991, 207-230.
Freeman, C. (1973), 'Malthus with a Computer', in Cole, H.S.D. *et al.* (eds.), *Models of Doom: A Critique of The Limits to Growth*, New York, Universe Books, 1973, 5-13.
___ (1990), 'Schumpeter's *Business Cycles* Revisited', in Heertje and Perlman (1990), 17-38 (also in Witt, 1993, 40-61).
___ (1992), 'Innovation, Changes of Techno-Economic Paradigm and Biological Analogies in Economics', in Freeman, C., *The Economics of Hope: Essays on Technical Change, Economic Growth and the Environment*, Pinter, London and New York, 1992, 121-142.
___, Clark, J. and Soete, L. (1982), *Unemployment and Technical Innovation: A Study of Long Waves and Economic Development*, Pinter, London.
___ and Perez, C. (1988), 'Structural Crises of Adjustment, Business Cycles and Investment Behaviour', in Dosi *et al.* (1988), 38-66.
Freeman, M. and Aldcroft, D.H. (1985), *The Atlas of British Railway History*, Croom Helm, London.
Friedman, D. (1991), 'Evolutionary Games in Economics', *Econometrica*, Vol. 59, 637-666.
Friedman, M. (1953), 'The Methodology of Positive Economics', in Friedman, M., *Essays in Positive Economics*, University of Chicago Press, Chicago, 1953, 3-43.
Galbraith, J.K. (1952/72), *American Capitalism*, Penguin, Harmondsworth, 1972.

Georgescu-Roegen, N. (1971), *The Entropy Law and the Economic Process*, Harvard University Press, Cambridge, Mass.
Gerybadze, A. (1982), *Innovation, Wettbewerb und Evolution: Eine mikro- und mesoökonomische Untersuchung des Anpassungsprozesses von Herstellern und Anwendern neuer Produzentengüter*, J.C.B. Mohr (Paul Siebeck), Tübingen.
Gleick, J. (1988), *Chaos: Making of a New Science*, Cardinal, Penguin, London.
Goldstein, J.S. (1988), *Long Cycles: Prosperity and War in the Modern Age*, Yale University Press, New Haven and London.
Goodwin, R.M. (1951/82), 'Iteration, Automatic Computers, and Economic Dynamics', in Goodwin (1982), 99–107.
___ (1982), *Essays in Economic Dynamics*, Macmillan, London and Basingstoke.
___ (1986), 'The M-K-S System: The Functioning and Evolution of Capitalism', in Wagener and Drukker (1986), 14-21.
___ (1988), Walras and Schumpeter: The Vision Reaffirmed, Conference of the International Schumpeter Society, Siena, Italy, May 24-27, 1988.
___ (1990a), *Chaotic Economic Dynamics*, Clarendon, Oxford.
___ (1990b), 'Walras and Schumpeter: The Vision Reaffirmed', in Heertje and Perlman (1990), 39-49.
Gould, S.J. (1977), *Ontogeny and Phylogeny*, Harvard University Press, Cambridge, Mass.
___ and Eldredge, N. (1977), 'Punctuated Equilibria: The Tempo and Mode of Evolution Reconsidered', *Paleobiology*, Vol. 3, 115-151.
Haberman, R. (1977), *Mathematical Models: Mechanical Vibrations, Population Dynamics, and Traffic Flow*, Prentice-Hall, Englewood Cliffs, N.J.
Hannan, M.T. and Freeman, J. (1989), *Organizational Ecology*, Harvard University Press, Cambridge, Mass. and London.
Hanusch, H. (ed.) (1988a), *Evolutionary Economics: Applications of Schumpeter's Ideas*, Cambridge University Press, Cambridge.
___ (1988b), 'Introduction', in Hanusch (1988a), 1-7.
Hardin, G. (1960/75), 'The Competitive Exclusion Principle', in Whittaker, R.H. and Levin, S.A. (eds.), *Niche: Theory and Application*, Dowden, Hutchinson and Ross, Stroudsburg, Pa., 1975, 35-40.
Hardin, R. (1990), 'The Social Evolution of Cooperation', in Cook, K.S. and Levi, M. (eds.), *The Limits of Rationality*, University of Chicago Press, Chicago and London, 1990, 358-378.
Haugeland, J. (1985/89), *Artificial Intelligence: The Very Idea*, MIT Press, Cambridge, Mass. and London.
Hayek, F.v. (1948), 'Economics and Knowledge' and 'The Use of Knowledge in Society', in Hayek, F.v., *Individualism and Economic Order*, University of Chicago Press, Chicago, 1948.
Heertje, A. and Perlman, M. (eds.) (1990), *Evolving Technology and Market Structure: Studies in Schumpeterian Economics*, University of Michigan Press, Ann Arbor, Mich.
Heiner, R.A. (1983), 'The Origin of Predictable Behaviour', *American Economic Review*, Vol. 73, 560-595.
___ (1988), 'Imperfect Decisions and Routinized Production: Implications for Evolutionary Modelling and Inertial Technical Change', in Dosi *et al.* (1988), 148-169.
___ (1989), 'The Origin of Predictable Dynamic Behavior', *Journal of Economic Behavior and Organization*, Vol. 12, 233-257.
Hicks, J. (1969), *A Theory of Economic History*, Clarendon, Oxford.
Hirschleifer, J. (1977), 'Economics from a Biological Viewpoint', *Journal of Law and Economics*, Vol. 20, 1-52.
___ (1982/93), 'Evolutionary Models in Economics and Law: Cooperation versus Conflict Strategies', in Witt (1993), 195-254.
___ (1987), *Economic Behaviour in Adversity*, Wheatsheaf, Brighton.
___ and Coll, J.C.M. (1988), 'What Strategies Can Support the Evolutionary Emergence of Cooperation?', *Journal of Conflict Resolution*, Vol. 32, 367-398.

___ and ___ (1992), Selection, Mutation, and the Preservation of Diversity in Evolutionary Games, *Papers on Economics and Evolution*, No. 9202, ed. by the European Study Group for Evolutionary Economics.
Hirschman, A.O. (1958), *The Strategy of Economic Development*, Yale University Press, New Haven, Conn.
___ (1987), 'Linkages', in Eatwell *et al.* (1987), Vol. 3, 206-211.
Hodgson, G.M. (1989), 'Institutional Economic Theory: the Old Versus the New', *Review of Political Economy*, Vol. 1, 249-269.
___ (1993), *Economics and Evolution: Bringing Back Life Into Economics*, Polity, Cambridge.
___ and Screpanti, E. (eds.) (1991), *Rethinking Economics: Markets, Technology and Economic Evolution*, Elgar, Aldershot.
Hoffmeister, F. and Bäck, T. (1991), 'Genetic Algorithms and Evolution Strategies: Similarities and Differences', *Papers on Economics and Evolution*, No. 9103, ed. by the European Study Group for Evolutionary Economics.
Hofstadter, D.R. (1985a), 'Mathematical Chaos and Strange Attractors', in Hofstadter, D.R., *Metamagical Themas: Questing for the Essence of Mind and Pattern*, Penguin, Harmondsworth, 1985, 364-395.
___ (1985b), 'The Prisoner's Dilemma Computer Tournaments and the Evolution of Cooperation', in Hofstadter, D.R., *Metamagical Themas: Questing for the Essence of Mind and Pattern*, Penguin, Harmondsworth, 1985, 715-734.
Holland, J.H. (1975), *Adaptation in Natural and Artificial Systems*, University of Michigan Press, Ann Arbor, Mich.
___, Holyoak, K.J., Nisbett, R.E. and Thagard, P.R. (1986), *Induction: Processes of Inference, Learning, and Discovery*, MIT Press, Cambridge, Mass. and London.
___ and Miller, J.H. (1991), 'Artificial Adaptive Agents in Economic Theory', *American Economic Review. Papers and Proceedings*, Vol. 81, 365-370.
Hull, D.L. (1973), *Darwin and his Critics: The Reception of Darwin's Theory of Evolution by the Scientific Community*, University of Chicago Press, Chicago and London.
___ (1974), *Philosophy of Biological Science*, Prentice-Hall, Englewood Cliffs, N.J.
Iwai, K. (1984), 'Schumpeterian Dynamics, Part II: Technological Progress, Firm Growth and "Economic Selection"', *Journal of Economic Behavior and Organization*, Vol. 5, 321-351.
___ (1984/93), 'Schumpeterian Dynamics: An Evolutionary Model of Innovation and Imitation', in Witt (1993), 125-156.
Jacob, F. (1981/85), *Mulighedernes spil: Om det levendes mangfoldighed*, Hekla, n.p., 1985.
Jensen, K. and Wirth, N. (1974), *The Programming Language PASCAL: Revised Report*, Springer, New York.
Jensen, P.E. (1988), 'Joseph A. Schumpeter (1883-1950)', in Sørensen, H. and Fivelsdal, E. (eds.), *Fra Marx til Habermas: Samfundsudvikling og offentlig regulering*, Nyt fra Samfundsvidenskaberne, Copenhagen, 1988, 95-112.
Kaldor, N. (1961), 'Capital Accumulation and Economic Growth', in Lutz, F.A. and Hague, D.L. (eds.), *The Theory of Capital*, Macmillan, London, 1961, 177–222.
Kamien, M.I. and Schwarz, N.L. (1982), *Market Structure and Innovation*, Cambridge University Press, Cambridge.
Karp, R.M. (1986), 'Combinatorics, Complexity, and Randomness: Turing Award Lecture', *Communications of the ACM*, Vol. 29, 98-109.
Khalil, E.L. (1992), 'Economics and Biology: Eight Areas of Research', *Methodus*, Vol. 4, 29-45.
Khan, M.S. (1957), *Schumpeter's Theory of Capitalist Development*, Muslim University, Aligarh, India.
Kingsland, S.E. (1985), *Modeling Nature: Episodes in the History of Population Ecology*, University of Chicago Press, Chicago and London.
Knuth, D.E. (1968/73), *The Art of Computer Programming: Fundamental Algorithms*, 2nd edn., Addison-Wesley, Reading, Mass.

Koçak, H. (1986), *Differential and Difference Equations through Computer Experiments*, Springer, New York.
Kreps, D.M. (1990a), *Game Theory and Economic Modelling*, Clarendon, Oxford.
___ (1990b), *A Course in Microeconomic Theory*, Harvester Wheatsheaf, New York.
Kuhn, T.S. (1962/70), *The Structure of Scientific Revolutions*, 2nd edn., University of Chicago Press, Chicago and London.
Küppers, B.-O. (1986/90), *Der Ursprung biologischer Information*, 2nd edn., Piper, München and Zürich (also as: Information and the Origin of Life, MIT Press, Cambridge, Mass., 1990).
Kuznets, S. (1930/67), *Secular Movements in Production and Prices: Their Nature and Their Bearing Upon Cyclical Fluctuations*, Augustus M. Kelley, New York, 1967.
Lakatos, I. (1970), 'Falsification and the Methodology of Scientific Research Programmes', in Lakatos, I. and Musgrave, A. (eds.), *Criticism and the Growth of Knowledge,* Vol. 4 of Proceedings of the International Colloquium in the Philosophy of Science, Cambridge University Press, Cambridge, 1970, 91–195.
Langlois, R.N. (1989), 'What was Wrong With the Old Institutional Economics (and What is Still Wrong With the New)', *Review of Political Economy*, Vol. 1, 270-298.
Lawler, E.L., Lenstra, J.K., Kan, A.H.G.R. and Shmoys, D.B. (eds.) (1985), *The Travelling Salesman Problem: A Guided Tour of Combinatorial Optimization*, John Wiley & Sons, Chichester.
Lewis, H.R. and Papadimitriou, C.H. (1981), *Elements of the Theory of Computation*, Prentice-Hall, Englewood Cliffs, N.J.
Liebowitz, S.J. and Margolis, S.E. (1990), 'The Fable of the Keys', *Journal of Law and Economics*, Vol. 33, 1-25.
Lindgren, K. (1990), Evolution in a Population of Mutating Strategies, NORDITA Preprint 90/22 S, Copenhagen.
___ (1991), 'Evolutionary Phenomena in Simple Dynamics', in Langton, C.G., Taylor, Farmer and Rasmussen (eds.) (1991), *Artificial Life II,* Santa Fe Institute Studies in the Sciences of Complexity, Vol. 10, Addison-Wesley, Redwood City, Calif., 295-312.
Lloyd, C., Rapport, D. and Turner, J.E. (1975), 'The Market Adaptation of the Firm', in Day and Groves (1975), 119-135.
Lomborg, B. (1993), 'Cooperation in the Iterated Prisoner's Dilemma', *Papers on Economics and Evolution*, No. 9302, ed. by the European Study Group for Evolutionary Economics.
Lundvall, B.-Å. (ed.) (1992), *National Systems of Innovation: Towards a Theory of Innovation and Interactive Learning*, Pinter, London.
MacArthur, R.H. and Wilson, E.O. (1967), *The Theory of Island Biogeography*, Princeton University Press, Princeton, N.J.
Machlup, F. (1951/91), 'Schumpeter's Economic Methodology', in Wood (1991), Vol. 1, 232-242.
Marengo, L. (1992), 'Coordination and Organizational Learning in the Firm', *Journal of Evolutionary Economics*, Vol. 2, 313-326.
Marks, R.E. (1992), 'Breeding Hybrid Strategies: Optimal Behaviour for Oligopolists', *Journal of Evolutionary Economics*, Vol. 2, 17-38.
Marris, R. (1991), *Reconstructing Keynesian Economics with Imperfect Competition: A Desk-Top Simulation*, Elgar, Aldershot.
Marshall, A. (1890/1961), *Principles of Economics*, 9th (Variorum) ed. with annotations by C. W. Guillebaud, Macmillan, London, 1961.
___ (1898), 'Distribution and Exchange', *Economic Journal*, Vol. 8, 37-59.
___ (1919), *Industry and Trade: A Study of Industrial Technique and Business Organization; and their Influences on the Conditions of Various Classes and Nations*, 2 edn., Macmillan, London.
Marty, A.G. (1955), Analyse critique de l'oeuvre de Joseph Schumpeter, Thèse pour le Doctorat, Faculté de Droit de l'Université de Bordeaux.
Matthews, R.C.O. (1984/93), 'Darwinism and Economic Change', in Witt (1993), 159-185.

May, R.M. (ed.) (1976/81), *Theoretical Ecology: Principles and Applications*, Sinauer Associates, Sunderland, Mass., 1981.
Mayr, E. (1976), *Evolution and the Diversity of Life: Selected Essays*, Belknap Press of Harvard University Press, Cambridge, Mass. and London.
___ (1982), *The Growth of Biological Thought: Diversty, Evolution, and Inheritance*, Belknap Press of Harvard University Press, Cambridge, Mass. and London.
___ and Provine, W.B. (eds.) (1980), *The Evolutionary Synthesis: Perspectives on the Unification of Biology*, Harvard University Press, Cambridge, Mass. and London.
McCraw, T.K. (ed.) (1988), *The Essential Alfred Chandler: Essays Toward a Historical Theory of Big Business*, Harvard Business School Press, Boston, Mass.
McKelvey, M. (1991), 'How do National Systems of Innovation Differ?: A Critical Analysis of Porter, Freeman, Lundvall and Nelson', in Hodgson and Screpanti (1991), 117-137.
Meadows, D.H., Meadows, D.L., Randers, J. and Behrens, W.W. (1972), *The Limits to Growth*, Universe Books, New York.
Menger, C. (1871/1981), *Principles of Economics*, New York University Press, New York and London, 1981.
___ (1883/1985), *Investigations into the Method of the Social Sciences with Special Reference to Economics*, New York University Press, New York, 1985.
Metcalfe, J.S. (1988), 'The Diffusion of Innovation: An Interpretative Survey', in Dosi *et al.* (1988), 560-589.
Milgate, M. (1987), 'Goods and commodities', in Eatwell *et al.* (1987), Vol. 2, 546-549.
Milgrom, P.R., North, D.C. and Weingast, B.R. (1991), 'The Role of Institutions in the Revival of Trade: The Law Merchant, Private Judges, and the Champaigne Fairs', *Economics and Politics*, Vol. 2, 1-23.
Mill, J.S. (1848), *Principles of Political Economy with some of their Applications to Social Philosophy*, John W. Parker, London.
Mirowski, P. (1983), 'An Evolutionary Theory of Economic Change: A Review Article', *Journal of Economic Issues*, Vol. 17, 757-768.
Mokyr, J. (1990), *The Lever of Riches: Technological Creativity and Economic Progress*, Oxford University Press, New York and Oxford.
Moos, L.S. (1990), 'Evolutionary Change and Marshall's Abandoned Second Volume', *Économie Appliquée*, Vol. 43, 85-98.
Mosekilde, E. and Larsen, E.R. (1988), 'Deterministic Chaos in the Beer Production-Distribution Model', *System Dynamics Review*, Vol. 4, 131-147.
Mowery, D. and Rosenberg, N. (1979/82), 'The Influence of Market Demand upon Innovation: A Critical Review of some Recent Empirical Studies', in Rosenberg (1982), 193-241.
Nelson, R.R. (1959a), 'The Economics of Invention: A Survey of the Literature', *Journal of Business*, Vol. 32, 101-127.
___ (1959b), 'The Simple Economics of Basic Scientific Research', *Journal of Political Economy*, Vol. 67, 297-306.
___ (1968), 'A "Diffision Model" of International Productivity Differences in Manufacturing Industry', *American Economic Review*, Vol. 58, 1219-1248.
___ (1987), *Understanding Technical Change as an Evolutionary Process*, North-Holland, Amsterdam.
___ (ed.) (1993), *National Innovation Systems: A Comparative Study*, Oxford University Press.
___ and Winter, S.G. (1973), 'Toward an Evolutionary Theory of Economic Capabilities', *American Economic Review*, Vol. 63, 440-449.
___ and ___ (1974), 'Neoclassical vs Evolutionary Theories of Economic Growth: Critique and Prospectus', *Economic Journal*, Vol. 84, 886-905.
___ and ___ (1978), 'Forces Generating and Limiting Concentration under Schumpeterian Competition', *Bell Journal of Economics*, Vol. 9, 524-548.
___ and ___ (1982), *An Evolutionary Theory of Economic Change*, Belknap Press of Harvard University Press, Cambridge, Mass. and London.

___, ___ and Schuette, H.L. (1976), 'Technical Change in an Evolutionary Model', *Quarterly Journal of Economics*, Vol. 90, 90-118.
Newell, A. and Simon, H.A. (1976), 'Computer Science as Empirical Inquiry; Symbols and Search: Turing Award Lecture', *Communications of the ACM*, Vol. 19, 113-126.
Nørretranders, T. (1991), *Mærk verden: En beretning om bevidsthed*, Gyldendal, Copenhagen.
Oakley, A. (1990), *Schumpeter's Theory of Capitalist Motion: A Critical Exposition and Reassessment*, Elgar, Aldershot.
O'Brien, P. (1977), *The New Economic History of the Railways*, Croom Helm, London.
O'Driscoll, G.P.J. (1986), 'Money: Menger's Evolutionary Theory', *History of Political Economy*, Vol. 18, 601-616.
Parijs, P.V. (1981), *Evolutionary Explanation in the Social Sciences: An Emerging Paradigm*, Rowman and Littlefield, Totowa, N.J.
Pearl, R. and Reed, L.J. (1920), 'On the Rate of Growth of the Population of the United States since 1790 and its Mathematical Representation', *Proceedings of the National Academy of Sciences*, Vol. 6, 275-288.
Penrose, E.T. (1952), 'Biological Analogies in the Theory of the Firm', *American Economic Review*, Vol. 42, 804-819.
___ (1959/80), *The Theory of the Growth of the Firm*, Basil Blackwell, Oxford, 1980.
Perez, C. (1983), 'Structural Change and the Assimilation of New Technologies in the Economic and Social System', *Futures*, Vol. 15, 357-375.
Perroux, F. (1935/65), *La pensée économique de Joseph Schumpeter: Les dynamiques du capitalisme*, Droz, Genève, 1965.
___ (1950/69), 'Les espaces économiques', in Perroux (1969), 159-177.
___ (1955/69), 'La notion de pôle de croissance', in Perroux (1969), 178-190.
___ (1969), *L'Économie du XXe siècle*, 3rd edn., Presses Universitaires de France, Paris.
Pianka, E.R. (1983), *Evolutionary Ecology*, 3rd edn., Harper & Row, New York.
Polya, G. (1945/71), *How to Solve It: A New Aspect of Mathematical Method*, 2nd edn., Princeton University Press, Princeton, N.J., 1971.
Popper, K.R. (1972), *Objective Knowledge: An Evolutionary Approach*, Clarendon, Oxford.
___ (1974/82), *Unended Quest: An Intellectual Autobiography*, Fontana/Collins, n.p., 1982.
Provine, W.B. (1971), *The Origins of Theoretical Population Genetics*, University of Chicago Press, Chicago and London.
Rayward-Smith, V.J. (1986), *A First Cource in Computability*, Blackwell, Oxford.
Rich, E. (1983), *Artificial Intelligence*, McGraw-Hill, Singapore.
Rogers, E.M. (1962/83), *Diffusion of Innovations*, 3rd edn., Free Press, New York, 1983.
Rosenberg, N. (1982), *Inside the Black Box: Technology and Economics*, Cambridge University Press, Cambridge.
Rosser, J.B. (1992), 'The Dialogue between the Economic and the Ecological Theories of Evolution', *Journal of Economic Behaviour and Organization*, Vol. 17, 195-215.
Roughgarden, J. (1979), *Theory of Population Genetics and Evolutionary Ecology: An Introduction*, Macmillan, New York and London.
Sahal, D. (1985), 'Technological Guideposts and Innovation Avenues', *Research Policy*, Vol. 14, 61-82.
Samuelson, P.A. (1951), 'Schumpeter as a Teacher and Economic Theorist', in Harris, S.E. (ed.), *Schumpeter: Social Scientist*, Harvard University Press, Cambridge, Mass., 1951, 48-53.
___ (1981), 'Schumpeter as an Economic Theorist', in Frisch, H. (ed.), *Schumpeterian Economics*, Praeger, Eastbourne and New York, 1981, 1–27.
Sanderson, S.K. (1990), *Social Evolutionism: A Critical History*, Basil Blackwell, Cambridge, Mass. and Oxford.

Saviotti, P.P. (1988), A Characteristics Approach to Technological Evolution and Competition, mimeo, University of Manchester.
___ and Metcalfe, J.S. (eds.) (1991), *Evolutionary Theories of Economic and Technological Change: Present Status and Future Prospects*, Harwood, London.
Schneider, E. (1970), *Joseph A. Schumpeter: Leben und Werk eines grossen Sozialökonom*, J.C.B. Mohr (Paul Siebeck), Tübingen.
Schotter, A. (1981), *The Economic Theory of Social Institutions*, Cambridge University Press, Cambridge.
Schuette, H.L. (1980), The Role of Firm Financial Rules and a Simple Capital Market in an Evolutionary Model of Industry Growth, Doctoral dissertation, University of Michigan.
Schumpeter, J.A. (1906/52), 'Über die mathematische Methode der theoretischen Ökonomie', in Schumpeter (1952), 529–548.
___ (1908), *Das Wesen und der Hauptinhalt der theoretischen Nationalökonomie*, Duncker & Humblot, Leipzig.
___ (1909/52), 'Bemerkungen über das Zurechnungsproblem', in Schumpeter (1952), 266-319.
___ (1910), 'Über das Wesen der Wirtschaftskrisen', *Zeitschrift für Volkswirtschaft, Sozialpolitik und Verwaltung*, Vol. 19, 271-325 (repr. in Schumpeter, 1987, 227-274).
___ (1912), *Theorie der wirtschaftlichen Entwicklung*, Duncker & Humblot, Leipzig.
___ (1912/26), *Theorie der wirtschaftlichen Entwicklung: Eine Untersuchung über Unternehmergewinn, Kapital, Kredit, Zins und den Konjunkturzyklus*, 2nd. rev. edn., Duncker & Humblot, Munich and Leipzig, 1926.
___ (1912/34), *The Theory of Economic Development: An Inquiry into Profits, Capital, Credit, Interest and the Business Cycle*, Oxford University Press, London, 1934.
___ (1928/51), 'The Instability of Capitalism', in Schumpeter (1951a), 47-72 (also in Witt, 1993, 14-39).
___ (1933/51), 'The Common Sense of Econometrics', in Schumpeter (1951a), 100-107.
___ (1937/51), 'Preface to Japanese Edition of "Theorie der wirtschaftlichen Entwicklung"', in Schumpeter (1951a), 158-163.
___ (1939), *Business Cycles: A Theoretical, Historical, and Statistical Analysis of the Capitalist Process*, 2 vols., McGraw–Hill, New York and London.
___ (1941/91), 'An Economic Interpretation of Our Time: The Lowell Lectures', in Schumpeter (1991b), 339-400.
___ (1942/87), *Capitalism, Socialism and Democracy*, Introduction by T. Bottomore, Unwin, London, 1987.
___ (1949/51a), 'The Historical Approach to the Analysis of Business Cycles', in Schumpeter (1951a), 308-315.
___ (1951a), *Essays on Economic Topics*, ed. R.V. Clemence, Kennikat, Port Washington, N.Y.
___ (1951b), *Ten Great Economists: From Marx to Keynes*, 1969 edn., Foreward by E. B. Schumpeter, Oxford University Press, New York.
___ (1952), *Aufsätze zur ökonomischen Theorie*, eds. E. Schneider and A. Spiethoff, J.C.B. Mohr (Paul Siebeck), Tübingen.
___ (1954), *History of Economic Analysis*, ed. E. B. Schumpeter, George Allen & Unwin, London.
___ (1987), *Beiträge zur Sozialökonomik*, ed. S. Böhm, Böhlau, Vienna.
___ (1991a), 'Letters by Schumpeter', in Swedberg (1991), 209-238.
___ (1991b), *The Economics and Sociology of Capitalism*, ed. R. Swedberg, Princeton University Press, Princeton, N.J.
Shackle, G.L.S. (1965), *A Scheme of Economic Theory*, Cambridge University Press, Cambridge.
___ (1967), *The Years of High Theory: Invention and Tradition in Economic Thought 1926-1939*, Cambridge University Press, Cambridge.
Shionoya, Y. (1990), 'Instrumentalism in Schumpeter's Economic Methodology', *History of Political Economy*, Vol. 22, 187-222.

Silverberg, G. (1988), 'Modelling Economic Dynamics and Technical Change: Mathematical Approaches to Self-Organization and Evolution', in Dosi et al. (1988), 531-559.
___, Dosi, G. and Orsenigo, L. (1988), 'Innovation, Diversity and Diffusion: A Self-Organization Model', *Economic Journal*, Vol. 98, 1032-1054.
Simmons, J. (1978), *The Railway in England and Wales 1830-1914*, Vol. 1, Leicester University Press.
Simon, H.A. (1947/65), *Administrative Behaviour: A Study of Decision-Making Process in Administrative Organization*, 2nd edn., Free Press, New York, 1965.
___ (1956/92), 'Rational Choice and the Structure of the Environment', in Egidi and Marris (1992), 39-54.
___ (1976/82), 'From Substantive to Procedural Rationality', in Simon (1982), Vol. 2, 424-443.
___ (1978/82), 'Rationality as Process and as Product of Thought', in Simon (1982), Vol. 2, 444-459.
___ (1981), *The Sciences of the Artificial*, 2nd edn., MIT Press, Cambridge, Mass. and London.
___ (1982), *Models of Bounded Rationality*, 2 vols., MIT Press, Cambridge, Mass. and London.
___ (1983), *Rationality in Human Affairs*, Blackwell, Oxford.
___ (1986), 'On the Behavioural and Rational Foundations of Economic Dynamics', in Day and Eliasson (1986), 21-41.
___ (1991a), *Models of My Life*, Basic Books, New York.
___ (1991b), 'Organisations and Markets', *Journal of Economic Perspectives*, Vol. 5, 25-44.
___ and Newell, A. (1958), 'Heuristic Problem Solving: The Next Advance in Operations Research', *Operations Research*, Vol. 6.
Smallwood, D.E. and Conlisk, J. (1979), 'Product Quality in Markets Where Consumers are Imperfectly Informed', *Quarterly Journal of Economics*, Vol. 93, 1-23.
Smith, A. (1776/1922), *An Inquiry into the Nature and Causes of the Wealth of Nations*, Methuen, London, 1922.
Smith, J.M. (1982), *Evolution and the Theory of Games*, Cambridge University Press, Cambridge.
Smithies, A. (1951), 'Memorial: Joseph Alois Schumpeter, 1883-1950', in Harris, S.E. (ed.), *Schumpeter: Social Scientist*, Harvard University Press, Cambridge, Mass., 1951, 11-23.
Stein, R.M. (1989), 'Real Artificial Life', *Byte*, Vol. 16, 289-298.
Stoneman, P. (1983), *The Economic Analysis of Technological Change*, Oxford University Press, Oxford.
Sugden, R. (1989/93), 'Spontaneous Order', in Witt (1993), 508-520.
Swedberg, R. (1991), *Schumpeter: A Biography*, Princeton University Press, Princeton, N.J.
Thomas, B. (1991), 'Alfred Marshall on Economic Biology', *Review of Political Economy*, Vol. 3, 1-14.
Thompson, F.M.L. (1976), 'Nineteenth Century Horse Sense', *Economic History Review*, Vol. 24.
Tinbergen, J. (1951/91), 'Schumpeter and Quantitative Research in Economics', in Wood (1991), Vol. 1, 175-179.
Tirole, J. (1988), *The Theory of Industrial Organization*, MIT Press, Cambridge, Mass. and London.
Tool, M.R. (ed.) (1988), *Evolutionary Economics*, 2 vols., M.E. Sharpe, Armonk, N.Y.
Universities-National Bureau of Economic Research (U-NBER) (1951), *Conference on Business Cycles*, National Bureau of Economic Research, New York.
Veblen, T. (1898/1961), 'Why is Economics not an Evolutionary Science?', in Veblen (1919/61), 56-81.

____ (1899/1961), 'The Preconceptions of Economic Science, I-III', in Veblen (1919/61).
____ (1919/61), *The Place of Science in Modern Civilisation and other Essays*, Russell & Russell, New York, 1961.
Wagener, H.-J. and Drukker, J.W. (eds.) (1986), *The Economic Law of Motion of Modern Society: A Marx—Keynes—Schumpeter Centennial*, Cambridge University Press, Cambridge.
Walras, L. (1874/1954), *Elements of Pure Economics or The Theory of Social Wealth*, transl. and ed. W. Jaffé, George Allen and Unwin, London, 1954.
Whittaker, R.H. and Levin, S.A. (eds.) (1975), *Niche: Theory and Application*, Dowden, Hutchinson and Ross, Stroudsburg, Pa.
Wieser, F.v. (1907/29), 'Arma virumque cano', in Wieser (1929), 335-345.
____ (1911/29), 'Das Wesen und der Hauptinhalt der theoretischen Nationalökonomie. Kritische Glossen', in Wieser (1929), 10-34.
____ (1914/27), *Social Economics*, translation of the 1924 German edn. of Wieser (1914), Adelphi, New York, 1927.
____ (1929), *Gesammelte Abhandlungen*, ed. F. A. von Hayek, J.C.B. Mohr (Paul Siebeck), Tübingen.
Williamson, O.E. (1985), *The Economic Institutions of Capitalism*, Free Press, New York.
Winter, S.G. (1964), 'Economic "Natural Selection" and the Theory of the Firm', *Yale Economic Essays*, Vol. 4, 225-272 (reprinted in Earl, 1988, 108-148).
____ (1971/93), 'Satisficing, Selection and the Innovating Remnant', in Witt (1993), 76-100.
____ (1975), 'Optimization and Evolution in the Theory of the Firm', in Day and Groves (1975), 73-118.
____ (1984/91), 'Schumpeterian Competition in Alternative Technological Regimes', in Wood (1991), 270-304.
____ (1986), 'Schumpeterian Competition in Alternative Technological Regimes', in Day and Eliasson (1986), 199-232.
____ (1987), 'Natural Selection and Evolution', in Eatwell *et al.* (1987), Vol. 3, 614-617.
____ (1991), 'On Coase, Competence, and the Corporation', in Williamson, O.E. and Winter, S.G. (eds.), *The Nature of the Firm: Origins, Evolution, and Development*, Oxford University Press, New York and Oxford, 1991, 179-195.
Wirth, N. (1982), *Programming in Modula-2*, Springer, Berlin.
Witt, U. (1985), 'Coordination of Individual Economic Activities as an Evolving Process of Self-Organization', *Economie Appliquée*, Vol. 37, 569-595.
____ (1987), *Individualistische Grundlagen der evolutorischen Ökonomik*, J.C.B. Mohr (Paul Siebeck), Tübingen.
____ (1991a), 'Reflections on the Present State of Evolutionary Economic Theory', in Hodgson and Screpanti (1991), 83-102.
____ (1991b),'Evolutionary Economics: An Interpretative Survey', *Papers on Economics and Evolution*, No. 9104, ed. by the European Study Group for Evolutionary Economics.
____ (ed.) (1992a), *Explaining Process and Change: Approaches to Evolutionary Economics*, University of Michigan Press.
____ (1992b), 'Turning Austrian Economics Into an Evolutionary Theory', in Caldwell and Boehm (eds.) (1992), *Austrian Economics: Tensions and New Directions*, Kluwer, Boston, 215-236.
____ (ed.) (1993), *Evolutionary Economics*, Elgar, Aldershot.
____ (1993b), 'Emergence and Dissemination of Innovations: Some Principles of Evolutionary Economics', in Day, R.H. and Chen, P. (eds.) (1993), *Nonlinear Dynamics and Evolutionary Economics*, Oxford University Press, New York and Oxford, 91-100.
Wood, J.C. (ed.) (1991), *J. A. Schumpeter: Critical Assessments*, 4 vols., Routledge, London and New York.
Young, A.A. (1928), 'Increasing Returns and Economic Progress', *Economic Journal*, Vol. 38, 527-542.

Index

Abelson, H. 199-200, 219-20
Abernathy, W.J. 58, 62, 220
adaptation 13, 21, 28, 42, 47, 53
Akerlof, G. 155, 220
Alchian, A.A. xii, 2, 9-13, 18, 24, 101-2, 133, 143-4, 186, 191, 194, 220
Aldcroft, D.H. 223
algorithms 96-8, 114, 138, 143
    notation 104, 109, 181, 198-9
    See also computer modelling, programming, variety-creation (genetic algorithms)
Allen, P.M. 24, 91, 220
Allen, R.L. 22, 220
analogies between biology and economics 5, 10, 12, 17, 20-1, 23-4, 80, 92-3, 189, 190-1
Andersen, E.S. xii, 27, 60, 74, 136, 220
Andersen, B. xiii
Annerstedt, J. xiii
Aoki, M. 182, 221
Arrow, K.J. 54-6, 58, 221
Arthur, W.B. 19, 52, 61, 144, 147-50, 181, 186, 221
Artificial Intelligence x, 7, 97-8, 100, 114, 138, 141, 143-5, 147, 164, 181, 200
Aumann, R.J. 18, 152, 221
Awan, A.A. 26, 195, 221
Axelrod, R. xi, 19, 24, 150-7, 161, 165-166, 174, 176-177, 179, 181-2, 184, 200-1, 221
Axelrod-type models
    computer tournaments 153-4, 166, 176, 182
    evolutionary stable strategy 18, 152, 154, 176, 179, 180
    noise 177, 184
    payoffs 151, 153-4, 177, 182, 184
    random numbers 153
    See also Prisoner's dilemma, strategies, Trader's Dilemma

Babbage, C. 49, 50
Bäck, T. 181, 225
Bagwell, P.S. 27, 221
Bain, J.S. x, 221
Batten, D.F. 156, 221

Baumol, W.J. ix, 21, 221
Behrens, W.W. 227
Berg, M. 30, 85, 221
biology *See* analogies, Darwinian analysis, density-dependence, ecological biology, ecological niches, evolutionary biology, Lamarckism
Blaug, M. 20, 221
Booker, L.B. 147, 221
Bottomore, T. 22, 221
Boulding, K.E. 24, 65, 94, 189, 221
Bowler, P.J. 191-3, 221
Boyd, R. 65, 222
Brookfield, H. 66-7, 222
business cycles 1, 3-4, 9, 22, 26, 28, 32-4, 37, 41, 43-4, 59-60, 81, 187
    and density-dependence 72-76
    Juglar cycles 74, 76, 93
    Kitchin cycles 93
    Kondratiev waves 46, 59, 74-6, 93

Casson, M. 94, 222
Cavalli-Sforza, L.L. 65, 222
Chamberlin, E.H. 61, 222
Chandler, A.D. ix, 59, 222
change, endogenous 1, 13, 21-22
chaos, deterministic 68, 76, 92
Char, B.W. 201, 219, 222
Chen, P. xi, 222
Church, A. 109
Clark, J. 223
Clark, K.B. 62, 220, 222
Clark, N. 23-4, 222
Clemence, R.V. 22, 222
Coll, J.C.M. 19, 224
Commons, J.R. 17
complexity 7-8, 19-20, 23, 31, 33, 45, 47-8, 51, 53, 55, 61, 64, 69, 70, 86, 91
computer modeling, analytical cycle 199-201
Conlisk, J. 124-6, 230
Cook, S.A. 97
Coombs, R. 139, 222
Cuvier, G, 27
Cyert, R.M. 100, 102, 145, 222

Dahl, O.-J. 107, 205, 219, 222
Dahmén, E. 19, 63, 65, 86, 222

Index   233

Darwin, C. 3, 27, 32, 79, 185, 189, 196, 197, 222
Darwinian analysis 10-2, 22-3, 143-4, 150, 190-3, 195-7
Dasgupta, P. 142, 222
David, P.A. 19, 21, 52, 61, 222
Davis, J. xiii
Davis, L. 147, 154, 222
Day, R.H. xi, 141, 222
density-dependence of diffusion and interaction 63-94, 169
  and innovation 77-8, 85, 89-90
  carrying capacity 28-9, 35, 37, 70-3, 77, 79, 86-9, 92
  coevolution 19, 78, 81, 151, 179
  coexistence 84-5
  competition 64, 71, 79, 81-6, 90
  crowding-type behaviour x, 71, 88, 90-1
  discrete or continuous time 92
  interdependence 63-4, 68, 78-87, 94
  maximum growth rate 70, 72, 77, 79, 86, 89, 92
  mutualism 64, 82-3, 85-6
  pioneering-type behaviour x, 71, 78-80, 88, 90-91
  predation 79, 82-3, 86
  predator-prey model 68, 79-80, 81, 83
  programming of 93
  See also diffusion, imitation, logistic curve, Volterra-Lotka equations
diffusion x, 19, 64-5, 67, 69, 71-8, 81, 89, 92
  See also density-dependence, logistic
Dijkstra, E.W. 222
Dion, D. 19, 151-2, 184, 221
division of knowledge 48, 61
division of labour 16, 48, 54, 61
Dobzhansky, T. 185
Doody, F.S. 22, 222
Dosi, G. 20, 23, 24, 45, 46, 47, 61, 103, 222, 230
Dreyfus, H.L. 141, 181, 223
Drukker, J.W. 231
Dvorak, A. 52

Earl, P.E. 95, 223
Eatwell, J. 64, 223
ecological biology 19, 27, 63-4, 90, 92
ecological niches 66, 85-6, 91-2, 94
economic history 21, 23
  automobilization 28, 30, 57, 59
  mail coaches 26-8, 32-3, 35, 77, 80-1, 83, 85-6
  QWERTY keyboard 19, 52, 55, 61
  railroadization x, 26-31, 36, 38, 59-60, 67, 72, 74, 77, 80-1, 86-8, 186, 188

economic history (Cont.)
  railways 26-35, 40, 59, 67-71, 74, 76-8, 80-1, 83, 85-6, 88-9, 90, 92
economic paradigms x, 19, 44, 45-49, 59
  and bounded rationality 46-49, 50
  and interactive learning 52-55
  and product life cycles 56-59
  and scientific paradigms 45
  and technological paradigms 45-7
  and technological trajectories 19, 47, 61
  and transaction costs 46, 49
  as interface specifications x, 45-59
  commodities in economic theory 61
  commodity abstraction 49-55, 59
  parameterised interface 47, 51-2
  versus radical innovation 45, 52, 56, 58-9
economics
  behavioural 100
  growth theory 139-40
  industrial x
  input-output analysis 66
  Keynesian analysis xii, 5, 17, 24, 74
  neoclassical economics 4, 6, 10, 15-7, 21, 56, 139, 151, 187, 193
  See also evolutionary economics, institutional economics, Marshallian analysis, Schumpeterian analysis, Walrasian analysis
Egidi, M. 223
Ekelund, R.B. 59, 223
Eldredge, N. 26, 193, 223-4
Eliasson, G. 141, 222
Elster, J. 13, 23, 191, 194-5, 197, 223
Emmeche, C. xiii, 184, 223
Ermoliev, Y.M. 221
evolutionary biology 3, 10, 14, 19, 21, 27, 45, 90, 186
  and functional explanation 191
  and orthogenetic explanation 191
  and teleological explanation 191
  Darwinian explanation 27, 192-3, 196
  diversity of explanations 190-3, 196
  Lamarckist explanation 191-2, 196
  punctuated equilibria 26-7, 43-4, 176, 178, 193
  saltationist explanation 192-3, 196
  sociobiology 24
  stasis 26, 176, 179, 193
  synthesis 21, 185, 187, 190, 193
  teleological explanation 196
  terminology 187, 193
evolutionary economics
  algorithmic approach ix-xi, 1-2, 5-9, 21-2, 109, 138
  and functionalist explanation 194
  and rational-choice explanation 194

evolutionary economics (*Cont.*)
  Artificial Economic Evolution ix-xi, 95, 104, 109, 181, 186-7
  assumptions, 15
  crowding out of evolutionary perspectives ix, 2-3, 5, 10, 16-7, 20, 23-4
  definitions of 13-6, 24-5
  diversity of explanations 15, 23, 193-6
  empirical orientation of 1, 2-5, 15, 20
  formal approach xi, 1-3, 5- 6, 16-7, 20-2
  gradualist explanation 22, 196-7
  heterogeneity principle 4, 20, 35
  historical approach 2-5, 23
  history of 16-21, 2-4
  mechanisms 14, 17-8, 20, 22
  money 16
  natural-selection explanation 195
  new ix-xi, 1-2, 5, 9, 18-21, 23, 25, 40, 63, 151, 186-7, 196
  old x, 1-2, 4-5, 9, 17-8, 20-2, 196
  policy implications xi-xii
  population perspective 1-2, 6, 9-13, 20, 77
  reinforcement explanation 194-5
  routine frequencies 31-6
  saltationist explanation 22, 27, 195-7
  split between diffusion and innovation studies 63-5
  stylised facts 4, 8, 29, 38, 67, 72, 74, 78
  synthesis ix, xi, 10, 13-5, 20, 63-4, 101, 186-7, 190, 196
  typological perspective 10
evolutionary process
  concept of 14, 18, 20, 25
  irreversibility 9, 13, 15, 22, 26, 45, 196
  lock-in 11, 15, 19
  ontogenetic evolution 139
  phylogenetic evolution 139
  population 14, 19, 64-5, 67-8, 70, 77, 79, 80-2, 92, 94
  self-transformation 1, 21
  stasis 12, 14, 37, 44, 150-1, 170, 176, 178
  See also routines, segregation, selection, transmission, variety-creation

Fagerberg, J. xiii
Feldman, M.V. 65, 222
Fisher, R.A. 173, 193
fitness 12, 36
  absolute 33, 34
  ex ante 33, 35, 36, 91
  observed 30, 32, 33, 34, 35, 36, 60, 91

fitness (*Cont.*)
  relative 34
  theoretical 30, 32, 35, 36, 60, 91, 151, 172
Flood, M. 155, 223
Flores, F. 138, 141, 223
Foray, D. xi, 223
Forrester, J.W. 23, 93, 223
Foster, J. 24-5, 223
Freeman, C. xi, xiii, 23, 46, 141, 190, 223
Freeman, J. 19, 65, 94, 224
Freeman, M. 223
Friedman, D. 18, 181-2, 223
Friedman, M. 12-4, 101, 144, 193-4, 223
Frisch, R. 5

Galbraith, J.K.133-4, 223
games
  analytical approach 151-2, 181
  Chicken 182
  coordination 152
  experimental approach 150-184
  iterated 18, 24
  repeated 18
  See also Prisoner's Dilemma, Trader's Dilemma
Geddes, K.O. 222
genetic algorithms *See* variety-creation
Georgescu–Roegen, N. 17, 224
Gerybadze, A. 124, 133, 139, 224
Gleick, J. 92, 224
Goldberg, D.E. 221
Goldstein, J.S. 31, 224
Gonnet, G.H. 222
Goodwin, R.M. xi, 2-3, 5-8, 68-9, 71-2, 74, 76-7, 92-3, 199, 224
Gould, S.J. 26, 139, 193, 224
Gresham, T. 50, 155
Groves, T. 222

Haberman, R. 94, 224
Haldane, J.B.S. 193
Hannan, M.T. 19, 65, 94, 224
Hanusch, H. 24, 224
Hardin, G. 84, 224
Hardin, R. 155, 224
Haugeland, J. 96, 98, 138, 145, 181, 224
Hayek, F.v. 22, 47, 48, 61, 138, 224
Hébert, R.F. 59, 223
Heertje, A. 224
Heiner, R.A. 101, 124, 138, 224
Hicks, J. 156, 224
Hirschleifer, J. 18, 24, 181, 224
Hirschman, A.O. 65-6, 225
Hoare, C.A.R. 222
Hodgson, G.M. 22-3, 25, 139, 225
Hoffmeister, F. 181, 225
Hofstadter, D.R. 92, 182, 225

Holland, J.H. 144-7, 150, 154, 155, 164, 174, 181, 183-4, 221, 225
Holyoak, K.J. 225
Hull, D.L. 196-7, 225
Huxley, T.H 185
Hyldtoft, O. xiii

imitation 12, 15, 33, 63, 77, 151, 172-3, 176
  See also diffusion, logistic
innovation 12, 15, 26-31, 33, 35, 40-6, 50, 56-9, 61, 64-81, 83, 85, 87, 90, 92, 151, 156, 172-6, 184
  incremental 28, 40, 56-7, 62, 65, 142
  macro 40
  micro 40
  product 52-5, 59
  radical 40, 58, 62, 65, 67
  swarms 65, 72-5, 80
  See also Nelson-and-Winter models, Schumpeterian analysis
institutional economics 4, 14, 17, 20, 22-5, 40
institutions 15-6, 18-9, 25, 50-1, 152, 182, 186
Iwai, K. 65, 77, 141, 225

Jacob, F. 61-2, 225
Jensen K. 198, 225
Jensen P.E. 4, 225
Johnson, B. xiii
Juma, C. 23-4, 222

Kaldor, N. 4, 225
Kamien, M.I. 142, 225
Kan, A.H.G.R. 226
Kaniovski, Y.M. 221
Karp, R.M. 97, 225
Keynes, J.M. 2, 4, 69
Khalil, E.L. 17, 225
Khan, M.S. 22, 225
Kingsland, S.E. 64, 68, 92, 225
knowledge 13-4, 45, 47-51, 54-5, 61, 71, 76, 99, 100, 103, 124, 138, 142, 189, 196
  tacit 129, 136
Knudsen, C. xiii
Knuth, D.E. 96, 225
Koçak, H. 92, 226
Køster, D. xiii
Kreps, D.M. 18, 138, 144, 152, 226
Kuhn, T.S. 45, 226
Küppers, B.-O. 144, 226
Kuznets, S. 64, 92, 226

Lakatos, I. 45, 132, 226
Lamarck, J.B.d. 191
Lamarckism 11, 143, 150, 191-2, 194
Langlois, R.N. 22, 226

Larsen, E.R. 69, 227
Lawler, E.L. 97, 226
learning 11, 15, 18, 52-4, 98, 106, 114, 146, 148, 172, 181
  by doing 54, 147
  by using 54, 58, 142
  about probabilities 125, 148, 149, 150, 181
  automata 147, 148
  confidence-building 148-9, 181
  interactive 52-55, 57, 142
  trial-and-error 6, 7, 10, 16, 143, 148
  See also rationality
Leibnitz, G.W.v. 197
Lenstra, J.K. 226
Leontief, W. 66
Levin, S.A. 94, 231
Lewis, H.R. 97, 226
Liebowitz, S.J. 61, 226
Lindgren, K. xiii, 176, 177-80, 184, 226
Linné, K.v. 197
Lloyd, C. 87, 94, 226
logistic
  curve x, 64, 67-8, 72, 74-7, 86-9, 92-3
  diffusion 75, 77
  equation 67-71, 74-9, 81-3, 86, 92
  and Gomperz curve 92
  See also density-dependence, imitation, Lotka-Volterra equations
Lomborg, B. 176, 179, 183-4, 226
Lotka, A.J. 68
Lotka-Volterra equations 78-9, 81-3, 86, 88, 92
Lundvall, B.-Å. xiii, 19, 66, 135-6 221, 226

MacArthur, R.H. 79, 90, 226
Machlup, F. 23, 226
Malthus, T.R. 23, 193
Manturana, H. 141
March, J.G. 100, 102, 145, 222
Marengo, L. 182, 226
Margolis, S.E. 61, 226
Markov process 18, 66, 107, 125, 127, 205, 209
Marks, R.E. 174, 182, 226
Marris, R. 200, 223, 226
Marshall, A. 3, 16-7, 21, 23, 59, 61, 196-7, 226
Marshallian analysis 1
  economies of scale 16
  representative firm 16, 17, 23
Marty, A.G. 22, 226
Marx, K. 16, 24, 69
Matthews, R.C.O. 194-5, 226
May, R.M. 79, 92, 227
Mayr, E. 10, 21, 185, 187, 191-3, 196-7, 227

McCraw, T.K. 59, 227
McKelvey, M. xiii, 19, 227
Meadows, D.H. 23, 227
Meadows, D.L. 227
Mendel, G.J. 193
Menger, C. 16, 17, 20, 227
Metcalfe, J.S. xi, xiii, 19, 23, 65, 227
methodology
    instrumentalist 12-4, 23, 193-4
    mechanism-oriented 13, 23, 189, 19-4
    Methodenstreit 4, 23
    punctualism 27
    realist 13-4, 23, 193, 196
Milgate, M. 61, 223, 227
Milgrom, P.R. 152, 227
Mill, J.S. 50, 227
Miller, J.H. 144, 147, 150, 225
Mirowski, P. 142, 227
Mitchell, W.C. 23
Mokyr, J. 40, 195, 227
Monagan, M.B. 222
Moos, L.S. 23, 227
Morgan, L.H. 24
Mosekilde, E. xiii, 69, 227
Mowery, D. 58, 227

National Bureau of Economic Research 3, 23, 230
Nelson, R.R. x-xi, 16, 18-20, 24-5, 55, 77, 95-6, 101-6, 109-11, 115, 118, 131-43, 150-1, 175, 198, 200-1, 205-6, 214-5, 223, 227
Nelson-and-Winter models xi, 95-6, 104, 105-24, 130-2, 134, 138-41, 181-2, 186
    bankruptcy 215-6
    capital accumulation 102, 104-5, 107-8, 112-113, 115, 120, 124, 130, 135, 141, 205, 216
    desired investment 104, 106-8, 112-3, 214
    finance 104, 106, 108, 112, 122-3, 139, 215
    imitation 104-8, 110-2, 115-20, 122-4, 127-31, 135-7, 141, 213
    indicators 216
    initialisation 205, 209-10
    innovation 101-2, 104-8, 110-2, 115-20, 122-3, 128, 135-7, 140-1, 213
    innovation and firm size 133-5
    mark-up 106, 112-3, 210, 215
    national systems of innovation 19, 131, 135, 137
    object-oriented approach 140-1, 216-8
    parameters 95, 107, 115, 119-20, 123, 126-7, 140, 205, 207
    process innovation 124, 128
    product innovation 124, 127-31

Nelson-and-Winter models (*Cont.*)
    productivity 102, 105-7, 110-2, 115-6, 120, 122-4, 128-9, 131, 134-6, 139-0
    programming of 18, 101, 107, 114-5, 201, 205-18
    random numbers 115, 117, 119, 120, 122, 127, 129, 141, 201, 203, 206-7, 209, 213
    satisficing mechanism 96-100, 102, 118-9, 123-7, 138-9, 141
    search space 100, 102-3, 105, 107, 115-6, 131, 133, 135-7, 139, 141, 205, 207-9
    short-run supply and demand 104-5, 108, 110-1, 205, 212
    simulation results 115-8
    state transformation 104, 108, 205, 209, 210-1
    technical change 101, 104, 107-8, 205, 212-3
    See also rationality, Schumpeterian analysis
Newell, A. 97-9, 138, 145, 228, 230
Newman, P. 223
Newton, I. 196
Nisbett, R.E. 225
Nørretranders, T. 138, 228
North, D.C. 227

O'Brien, P. 59, 228
O'Driscoll, G.P.J. 16, 228
Oakley, A. 22, 228
Orsenigo, L. 230
Østergaard, J. xiii, 93, 219

Papadimitriou, C.H. 97, 226
paradigms *See* economic paradigms
Parijs, P.V. 194-5, 228
Pavitt, K. xiii
Pearl, R. 64, 67, 92, 228
Pedersen, J.L. xiii
Penrose, E.T. 12, 101, 190, 191, 228
Perez, C. 46, 223, 228
Perlman, M. 224
Perroux, F. 22, 63, 65-7, 91, 228
Pianka, E.R. 65, 90, 94, 228
Polya, G. 98, 228
Popper, K.R. 17, 36, 55, 91, 196, 228
Post, E. 145
Prisoner's Dilemma, iterated xi, 19, 151-3, 155-6, 176-7, 181-3, 188
    See also Axelrod-type models, strategies, Trader's Dilemma
programming 95-8, 199-201
    Artificial Adaptive Agents 144-50
    classifier systems 146
    data-driven 145

programming (*Cont.*)
    modular 107
    recursion 205, 209
    rule-based systems xi, 145-7
    structured 107
    See also algorithms, computer simulations, programming languages
programming language 6, 7, 8
    DYNAMO/STELLA 93, 199
    FORTRAN 140, 201, 219
    LISP 114, 199-200, 219
    MAPLE 200-3, 217, 219
    PASCAL 198-201, 216, 219
Provine, W.B. 21, 185, 187, 193, 227-8

Randers, J. 227
Rapport, D. 226
rationality
    bounded 10, 18, 46, 48, 98-100, 102, 138, 144, 147, 151, 164
    heuristics 7, 30, 36, 47, 96, 97-9, 114, 123-4, 143-4, 147-9, 166
    optimisation 147
    unbounded 10, 98-9, 132, 138, 144, 151
Rayward-Smith, V.J. 97, 228
Rechenberg, I. 181
Reed, L.J. 64, 228
Ricardo, D. 61
Rich, E. 141, 145, 181, 228
Richerson, P.J. 65, 222
Rogers, E.M. 64, 228
Rosenberg, N. 54, 58, 227-8
Rosser, J.B. 17, 228
Roughgarden, J. 65, 79, 228
routines 13, 26-38, 40, 42-4, 47, 51, 63-6, 78-88, 90-1, 94
    classes of 35, 60
    system of 26-31, 33, 36-7, 43-4

Sahal, D. 52, 228
Samuelson, P.A. 2-5, 59, 228
Sanderson, S.K. 24, 228
Saviotti, P.P. xi, 23, 52, 222, 229
Schneider, E. 22, 229
Schotter, A. 22, 229
Schuette, H.L. 133, 140, 201, 228-9
Schumpeter, J.A. ix-xii, 1- 6, 9, 13, 16-7, 20-4, 26-7, 29-45, 53, 57-61, 63, 65, 67-9, 72, 74, 76-81, 85, 87-93, 95-6, 110, 119, 132-5, 144, 186-90, 195-6, 229
Schumpeterian analysis 4, 24, 25, 27, 31, 34, 42, 66, 69, 74, 102, 187, 190
    and vision 29, 60, 77, 80, 186
    banks 16, 35, 71-2, 78
    circular flow 29, 38, 40, 59

Schumpeterian analysis (*Cont.*)
    competition xi, 20, 95-6, 139, 186
    creative destruction ix, 134
    development blocks 19, 65, 86
    'dynamics' 40-1
    economic development 38, 40, 49
    economic evolution 31-2, 37-8, 40-1, 45, 60
    entrepreneur 16, 28-30, 33, 35-6, 41, 43-6, 49, 53, 55-9, 63, 72, 77, 89-91
    'growth' 31-4, 37, 39, 53, 68
    growth poles 19, 65-67, 91
    innovation 30-1, 34, 44
    innovativeness of large firms 133-4
    manager 29, 36, 44-6, 49, 55-6, 78, 89, 90
    Mark I and II models 119-20, 142
    missing invention 53
    reconstruction of 26-44
    'statics' 40-1, 60
    theses on cycles and evolution 40-44
    See also business cycles, economic paradigms, innovation, Nelson-and-Winter models, Schumpeterian scheme
Schumpeterian scheme, reconstruction of 26-7, 29-31, 36, 38-43, 60, 80
    α-operator 36-43, 60
    β-operator 36-8, 40, 42-3, 60
    Δ-state 38-9, 41-3, 60
    F-space 32-3, 36-9, 41-3
    G-set 32, 36-7, 38-9, 40-1
    irreversibility 39
    punctuated evolution x, 26-7, 36-41, 43-4
    Σ-state 38-9, 41-3, 60
Schwarz, N.L. 142, 225
Screpanti, E. 225
segregation, mechanism of 14
selection 12-3, 18, 34, 36, 44, 85-6, 89, 90
    and profits 11-3, 22
    classifier systems 146-8
    global optimum 12, 144, 148-50, 184, 191, 194
    local optima 15, 19, 144, 148, 150, 181, 192, 195
    mechanism of 11, 13-4, 37, 102
    natural 11-2, 18, 32, 102, 143, 189, 191-2, 194-6
    optimisation 13, 144, 146, 147-8, 174, 181
    units of 189-190, 196
Shackle, G.L.S. 13, 14, 23, 61, 229
Shaw, C. 97, 181

Shionoya, Y. 23, 229
Shmoys, D.B. 226
Silverberg, G. 65, 141, 223, 230
Simmons, J. 27, 83, 230
Simon, H.A. 13, 24, 61, 95, 97-100, 102, 138-9, 141, 143-4, 151, 180-1, 194, 228, 230
Smallwood, D.E. 124-6, 230
Smith, A. 16, 230
Smith, J.M. 18, 179, 184, 230
Smithies, A. 3, 230
Soete, L. 223
Spencer, H. 23-4, 32
Stein, R.M. 230
Stephenson, G. 83
Stephenson, R. 83
Stiglitz, J. 142, 222
Stoneman, P. 64, 230
strategies for the iterated Prisoner's Dilemma
   4-bit 163, 176, 178-9
   8-bit 163, 176, 178-9
   16-bit 179
   all-D 153-6, 161-3, 165, 177-8, 180
   all-P 155, 161-3, 177-8
   anti-tit-for-tat 163, 170, 177-8
   massive-retaliatory-strike 153, 167
   tit-for-tat 153-5, 161-3, 166, 170, 176-9, 182, 184
   tit-for-two-tats 163
   two-tits-for-one-tat 154,-5, 161-3
   See also Axelrod-type models, Prisoner's Dilemma, Trader's Dilemma
Sugden, R. 18, 182, 230
Sussman, G.J. 199, 200, 219-20
Swedberg, R. 3-4, 22-3, 230

Thagard, P.R. 225
Thomas, B. 16, 230
Thompson, F.M.L. 30, 230
Tinbergen, J. 6, 230
Tirole, J. x, 230
Tool, M.R. 24-5, 230
Townsend, P.L. 62, 220
Trader's Dilemma, iterated 49, 152, 155, 157-8, 161, 164, 166, 169, 171, 175-7, 180, 183-4
   discounting 172, 183
   evolutionary stable strategy 170
   noise 167, 183
   notary 157-60, 166-70, 175, 180, 183
   payoffs 156, 158, 161, 166-7, 169, 183
   See also Axelrod-type models, Prisoner's Dilemma, strategies
transmission 196
   heredity 12

transmission (*Cont.*)
   mechanism of 11, 14, 102
Turing, A. 97, 109, 145, 147
Turner, J.E. 226
Tylor, E.B. 24

Ungern-Sternberg 141
Utterback, J.M. 62, 220

variety-creation 196
   doubling 174-5, 177, 184
   genetic algorithms xi, 146-8, 154-5, 162, 164, 174-6, 181-4
   in biology 177
   inversion 147, 174-5
   mechanism of 14, 37, 102
   mutation 12, 143, 146-7, 171-2, 174-5, 177, 182-4
   recombination 146, 172-5, 182-3
   See also innovation
Veblen, T. 10, 17, 22-5, 230
Verhulst, P.F. 67, 68
Volterra, V. 68
Vries, H.d. 192

Wagener, H.-J. 231
Wallace, A.R. 79
Walras, L. 1, 6-9, 22, 42-3, 49, 187, 231
Walrasian analysis 1, 6, 8-9, 23, 42-3, 48-9, 56, 61
   algorithms 7-8
   disequilibrium 8
   entrepreneur 61
   servo-mechanisms 7-8, 23
   tâtonnement 6-8, 49, 159
Walsh, V. 222
Wangenheim, G.v. xiii
Watt, S.M. 222
Weingast, B.R. 227
Whittaker, R.H. 94, 231
Wiener, N. 7
Wieser, F.v. 195, 231
Williamson, O.E. 22, 46, 231
Wilson, E.B. 92
Wilson, E.O. 79, 90, 226
Winograd, T. 138, 141, 223
Winter, S.G. x-xii, 13, 16, 18, 20, 24-5, 30, 34, 60, 77, 91, 95-6, 101-6, 109-11, 115, 118, 120, 124, 131-44, 150-1, 175, 182-3, 186, 198, 200-1, 205, 206, 214-5, 227, 231
Wirth, N. 107, 198, 225, 231
Witt, U. xi-xiii, 21, 23, 60, 62, 64, 94, 231
Wood, J.C. xii, 231

Young, A.A. 16, 231